HAZARD CONTROL
POLICY IN BRITAIN

HAZARD CONTROL
POLICY IN BRITAIN

John C. Chicken

PERGAMON PRESS

Oxford · New York · Toronto
Sydney · Paris · Braunschweig

Pergamon Press Offices:

U. K. Pergamon Press Ltd., Headington Hill Hall, Oxford OX3 0BW, England

U. S. A. Pergamon Press Inc., Maxwell House, Fairview Park, Elmsford, New York 10523, U.S.A.

CANADA Pergamon of Canada, Ltd., 207 Queen's Quay West, Toronto 1, Canada

AUSTRALIA Pergamon Press (Aust.) Pty. Ltd., 19a Boundary Street, Rushcutters Bay, N.S.W. 2011, Australia

FRANCE Pergamon Press SARL, 24 rue des Ecoles, 75240 Paris, Cedex 05, France

WEST GERMANY Pergamon Press GMbH, 3300 Braunschweig, Postfach 2923, Burgplatz 1, West Germany

First edition 1975

Library of Congress Cataloging in Publication Data

Chicken, John C.
    Hazard control policy in Britain.

    Includes index.
    1. Industrial safety—Great Britain. I. Title.
HD7696.C5 1975   614.8'52'0941   75–12900
ISBN 0 08 019739 6

Printed in Great Britain by
A. Wheaton & Co., Exeter.

CONTENTS

v

FIGURES

PREFACE

The aim of this book is to examine the general nature of the
British hazard control policy making process, as it appeared at
the beginning of the nineteen seventies, before the
recommendations of the Robens' Committee on Safety and Health at
work were implemented. Attention is directed mainly at
identifying the apparent roles and goals of the various
organisations associated with policy making. This general
view of hazard control policy is built up from case studies of
areas the author has practical experience in. The case studies
cover: road transport, air transport, factories, nuclear power
reactors, and air contamination. These studies while not
including many specific hazards such as those associated with
shipping, food, explosives, and mining are sufficiently
representative of the whole spectrum of hazards that they give
a clear indication of the general nature of the hazard control
policy making process.

The analysis is divided into three main parts: the nature of
hazards, hazard control policy, and the role of interest groups.
Chapter 2 deals with the nature of hazards, the technical causes
and risks to life of the hazards associated with each activity
considered are outlined, and compared with the risks from
natural hazards. In Chapter 3 hazard control policy is examined
and a model of the policy making process is postulated, this is
followed by: a survey of current policy and the way it has
developed in the five activities studied, identification of
policy making centres and proximate policy makers, and
suggestions about the possible ways policy may develop in the
future. Interest groups are identified as one of the actors in
the policy making process, and in Chapter 4 their role is
examined in some detail. The examination of interest groups is
based on a review of the relevant literature and unique data
obtained from the answers of twenty seven interest groups to a
questionnaire.

The Author wishes to thank Professor Reynolds, Professor Nailor
and George Clark of Lancaster University for their encouragement
and help in pursuing the research on which this book is based.

Also the cooperation of the following bodies is acknowledged:
Society of Motor Manufacturers and Traders, Motor Industry
Research Association, Institution of Civil Engineers,
Institution of Mechanical Engineers, Royal Society for the
Prevention of Accidents, Automobile Association, Royal
Automobile Club, Motoring Which, Society of British Aerospace
Companies, Lloyds Aviation Underwriters Association, Royal
Aeronautical Society, Guild of Air Traffic Control Officers,
British Air Line Pilots Association, Air Registration Board,
Flight Safety Committee, Amalgamated Union of Engineering
Workers, Central Electricity Generating Board, Institution of
Professional Civil Servants, Confederation of British Industry
National Union of General and Municipal Workers, British Medical

Association, Association of Public Health Inspectors, National
Society for Clean Air, Clean Air Council, Civic Trust
Association of Municipal Corporations and the Conservative and
Unionist Central Office.

Bowdon

John C. Chicken

1975

CHAPTER 1

INTRODUCTION

This book examines the general nature of hazard control policy in Britain and the factors that appeared to influence its formation up to 1973, that is prior to the implementation of the Health and Safety at Work Act of 1974. The general view of hazard control policy is built up from case studies of five activities. The case studies cover road transport, air transport factories, nuclear power reactors and air contamination. These studies, while excluding many hazards such as those associated with shipping, food, explosives and mining, are sufficiently representative of the whole spectrum of technological hazards to give a clear indication of the general nature of the hazard control policy making process.

The starting point for this study is the assumption that the aim of responsible governments, such as we have in Britain, is to satisfy those needs of the nation that it perceives to exist, and which it is possible to satisfy. In pursuit of this aim a government has, as far as possible, to develop policies that are in harmony with current norms, and which resolve any demands for resources. These competing requirements arise in many ways and may be associated with any aspect of government, from maintaining law and order within the nation to defending the country's borders, and from the promotion of the economic development of the nation to providing social services that are relevant to the needs of the nation at a particular time. In developing a policy on a particular subject a government attempts to give the weight it considers appropriate to the views of the various parties that may be affected by the policy.

The British government system is considered to be a suitable system on which to base the study as it is an example of what Finer (1) calls a liberal-democratic type of government; which has the characteristic, relevant to this study, that it is sensitive to public opinion and allows such opinions to be overtly and freely expressed. In the context of the specialist policy field of hazard control the free expression of views may simply amount to proximate policy makers openly consulting organisations with specialist knowledge in the field. This process of consultation and the scope that it gives interest groups to influence policy is given particular attention.

In deciding which hazards and which interest groups to study a number of factors had to be considered. The range of hazards that result from man's activity in an advanced technological society, such as we have in Britain, is considerable. The spectrum of hazards considered are those that could loosely be called technological hazards, and excludes the hazards associated with war and violence. Technological hazards range

from the risk of being electrocuted by an electric iron to being
blown up by a defective gas heater and from the risk of being
killed in an accident involving a means of transport to being
killed by toxic substances released from an industrial process.
The problem was to select a sample of hazards that, besides
being representative of the whole range of technological hazards,
are well understood in the technical sense, meaning by this that
the nature and significance of the hazards are known. Also to
establish the extent of any influence of the organisational
pattern on hazards and hazard control policy each of the hazards
considered should, ideally, be representative of a particular
organisational pattern. An additional consideration taken into
account in deciding which hazards to study was the fact that
the author had, in the course of his career as a Chartered
Engineer, been concerned, to some extent, with road transport,
air transport, factories, nuclear power stations and air
contamination from industrial and domestic sources. In these
five fields the hazards are representative of the whole pattern
of technological hazards, and they are associated with different
organisational and control patterns, so they satisfied the
requirements for a representative sample. On this basis they
were selected as the sample of hazards to be studied. The
essential organisational patterns associated with the five
fields selected are as follows:-

1.   Road transport is essentially privately owned, privately
operated on publicly owned roads, and subject to detailed
government regulations that are somewhat difficult to enforce
universally because of the large number of people concerned.
Road transport, in its present form, based on the internal
combustion engine only really became significant from the
beginning of this century. The road system has a longer history,
but central government concern about developing a road system
only became important after the First World War.

2.   Air transport is partly privately owned and partly publicly
owned, and is subject to strict government control as to: which
routes can be used, the skill of crews, the reliability of the
aircraft, and the soundness of the aircraft's design. With only
a small number of operating organisations to be monitored strict
enforcement of control is a practical proposition. Air transport
has only been a viable proposition since the decade before the
Second World War, and the wartime development of aircraft for
military purposes accelerated the subsequent progress of civil
air transport.

3.   Factories are generally privately owned; under government
legislation they have to be registered and approved and are
subject to regular inspection by inspectors appointed by the
government. But because of the diversity and complexity of the
processes involved enforcement of uniform standards is difficult
with the present size of the inspectorate. Factory organisation
dates from the Industrial Revolution, and the legislation to
control factory hazards has its origin in the Health and Morals
of Apprentices Act 1802.

4.   Nuclear power reactors are operated by state owned
industries, such as the Central Electricity Generating Board and
the United Kingdom Atomic Energy Authority. They are by statute
subject to strict control procedures, which can be more readily
applied as there are only a small number of units to be
considered. Nuclear reactors are a recent innovation, being the
product of a technology that was developed during the Second
World War.

5.   Contamination of air by smoke is due to industrial processes
and domestic fires. The plants in which industrial processes
take place are partly privately owned and partly publicly owned.
By definition the domestic fires are all privately owned.
Responsibility for controlling this hazard is divided between
central and local government. Because of the large number of
units concerned and the diversity of control there has not been
a uniform reduction in this hazard throughout the country. The
problem of smoke is not new; it has been a matter of public
concern for centuries.

In the five fields considered it will be noticed that there is a
wide range of government involvement varying from direct and
complete control in the case of nuclear power reactors to divided
and somewhat indirect control in the case of air contamination.

Having selected the hazards to be studied the analysis of the
policy on their control was built up in four stages, which are
described in the chapters that follow. The first stage consists
of a description of the nature of hazards to show some of the
inherent problems in their control. In this description an
explanation of the technical causes of hazards is given, and to
put the hazards into perspective some statistics of their
frequency of occurrence are given. The probability of fatal
accidents occurring is used as an index of the seriousness of a
particular hazard, and this level of probability is compared
with that associated with natural hazards, such as receiving
lethal radiation from an exploding Super Novae.

In the second stage of the study the essential features of the
policy making process are identified, existing policy on hazard
control is described, and possible future policy developments
are indicated. This description of the policy making process is
constructed in five main steps. First, systems analysis
supplemented by a description of the probable roles and
interactions of the various actors is used to postulate a model
of the policy making process. Secondly, the technical and
institutional instruments that make up hazard control in the
five cases studied are outlined and the way they have developed
is illustrated. Thirdly, from the description of the development
of instruments of control the proximate policy makers and policy
making centres are identified. Fourthly, analysis of hazard
control policy is made, and this leads to an evaluation of the
model of the policy making process that was postulated. Finally,
possible future developments in hazard control policy are
suggested.

The third stage of the study is an analysis of the role of
interest groups appear to have played in influencing hazard
control policy in the five fields considered. For the analysis
a sample of interest groups was selected that appeared concerned
to have their views taken into account in the formation of policy
on hazard control. From the outset it was recognised that
interest groups have various motives for their operations and
that they operate at various stages in the policy making process.
The motives of interest groups have been classified by Wootton
(2) as being economic, integrative, or cultural, and the level
at which they operate has similarly been classified under three
orders according to whether the groups exert their influence at
local, national or international level. The analysis is based
mainly on the answers that the selected interest groups gave to
a questionnaire and indicates: the significance of the technical
and financial resources, the extent of their interaction with
proximate policy makers, their procedures for developing policy
and their goals. A measure of the significance of the interest
groups is inferred from the influence that seven of the groups
studied appear to have had on the hazard control policy proposals
put forward by Lord Robens' Committee on Safety and Health at
Work.

In the fourth and final stage of the study a number of tentative
conclusions are drawn about: the form of the policy making
process related to hazard control, the possible developments in
hazard control policy, and the role, organisation, and influence
of interest groups. Also a modification to the Wootton method
of classifying interest groups is suggested with the aim of
improving the indication of the nature and organisation of the
groups classified by this method.

CHAPTER 2

THE NATURE OF HAZARDS

Before plunging into the analysis of the formation of hazard
control policy, and the role that interest groups play, the
nature of the hazards to be considered is described. The
description, which outlines the technical causes and the
statistical significance of the hazards, is aimed at drawing
attention to the limitations the nature of the hazard imposes
on the range of policy strategies the policy maker can employ
to control hazards.

The first step in describing hazards must be to specify the way
the term hazard is to be used. For the purpose of this study a
hazard is defined as a set of circumstances that introduced a
quantifiable element of danger into life. The importance of
specifying hazards in quantifiable terms has been stressed by
Farmer (3), who argued that to describe hazards* as being simply
credible or incredible is meaningless as there is no logical way
of differentiating between such terms. This line of argument
was developed further by Otway and Erdman (4) who also made the
point that as a new hazard is recognised it will intuitively be
compared by the public with existing hazards, and that such
comparison is most suitably made in quantitative terms.

There are three general features of hazards that Otway and
Erdman (4) describe that help to put hazards into perspective,
these features are as follows:-

1.   The circumstances that give rise to a hazard are generally
     just part of some activity that is otherwise beneficial
     and this benefit has to be balanced against risk.

2.   All people are not equally exposed to hazards, so concern
     about the control of a hazard will not be equally
     distributed.

3.   An acceptable level of hazard is a compromise between the
     level of hazard that will be accepted by the public and
     the amount of safety related expenditure and operating
     expenditure the authorities responsible for the activity
     are willing to accept.

This leads to the question of what is the level of hazard that
will be acceptable as the basis for compromise between the
allocation of resources to its reduction and the allocation of
resources to other purposes. In our ordinary everyday life we

---

* Farmer in fact uses the term risk to cover the circumstances covered by
  the term hazard in this book.

are subject to natural hazards such as: lightning, earthquakes, and cosmic radiation over which man has no control, at present. A probability of occurrence can be assigned to all these natural events, and it is postulated that if the probability of an artificial hazard is small compared with natural hazards it will be considered to be acceptable. The corollary of this postulate is that if a hazard has a probability similar to, or greater than a natural hazard, it would generally be considered unacceptable, and some action would be required to reduce it.

A complete comparison of the significance of particular hazards would involve comparison of financial loss, injury, and death. For the purpose of this study death is taken as giving a reasonable index of the harms the hazards in each field considered can cause. There is less ambiguity about statistics on death rates, than those associated with financial loss or injury.

Before examining the hazards in the five selected fields an attempt is made to quantify the significance of natural hazards.

The death of man is the ultimate consequence of all the natural and artificial hazards he is exposed to in the course of his life It follows from the postulation above that current life expectancy is regarded as normal, any improvement an advantage, but any significant deterioration in life expectancy is unacceptable. The life expectancy of each person is different and is compounded of many hereditary and environmental factors. For example, a European has a greater life expectancy than an Indian, and anyone living in the twentieth century expects a longer life than his ancestors in the first century. Based on the British life expectancy tables (5), Fig.1 has been prepared, and compares the probabilities of normal death with the risk of death from various man-made hazards and with natural hazards, such as drowning (4), lightning (4), and the probability of lethal radiation arriving on earth from an exploding Super Novae (6).

From Fig.1 it is possible to infer that unacceptable hazards are those which have a probability of causing death within a year greater than 10^{-3}, acceptable hazards are those with a probability of death within a year of less than 10^{-6} and if the hazard has a probability of between 10^{-3} and 10^{-6}, then it is expected that some steps would be taken to reduce the hazard to an acceptable level.

ROAD TRANSPORT

Of the five activities considered road transport is the one which causes the greatest sudden loss of life, and the greatest number of injuries; air contamination is probably the greatest killer considered, but its effect becomes apparent more gradually. During 1969, when there were 14,751,900 licenced road vehicles,

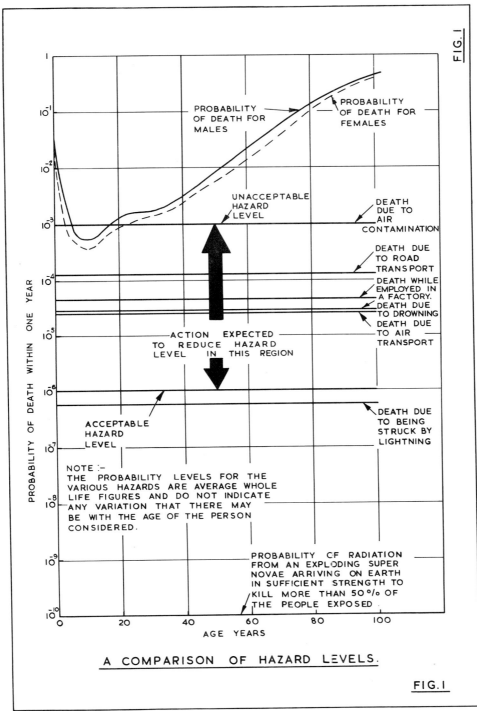

A COMPARISON OF HAZARD LEVELS.

FIG.I

7,383 people were killed on the roads, and 345,811 injured (7)*
By any standard this represents an awful carnage, and in more
dramatic terms is equivalent to injuring everyone in a major
city every year, or wiping out the whole population of a medium
sized town every ten years.

The figure of 7,383 people killed during 1969 is quite typical
of the death toll on the roads in recent years. If it is
considered that the risk is equally divided among the whole
population of the country, which in 1969 was 55,534,000 (8), it
suggests that in any one year the probability of being killed on
the roads is of the order of 1.33×10^{-4}. From earlier
arguments this is clearly a level of probability at which it is
considered some action should be taken to reduce the risk.

These death and injury rates, as will be shown later, resulted
in demands from interest groups for safer roads. There are also
indications that cost benefit techniques are now being used to
evaluate whether or not a particular safety provision such as
crash barriers, should be provided.**

Professor R.H. Macmillan, Director of the Motor Industry
Research Association, in his 1971 Motor Show lecture (9), put
the causes of road accidents in the following order of
importance:- human behaviour, road conditions and vehicle design
In this lecture it was stressed that even with perfect vehicles
a large proportion of accidents would still take place, and that
better roads and improvement in driving have the greatest
potential for accident reduction. Professor Macmillan also
suggested that there are four limitations which prevent the
vehicle designer from doing more than he does at present.

1.   Technical - lack of knowledge, or current capability.

2.   Economic - solution known but too costly.

3.   Styling - conflict with known or assumed customer
                preference.

4.   Legislation - conflict of national requirements.

A rather more quantitative view of the cause of road accidents
was given by the Safety Division of the Road Research Laboratory
of the Ministry of Transport in the Laboratory's Annual Report
(10) in which they drew attention to the following important
characteristics of road accidents:

 * The figures for 1972 show 7,779 deaths and 352,013 injured. so there has
been little change. The increase in the accident rate has been slightly
lower than the increase in the number of vehicles which in 1972 had increased
to 16,117,000. Ref. Annual Abstract of Statistics 1973 published HMSO London.

** This point is discussed further on page 33

1.   Casualty rates per vehicle - kilometer are higher in urban
     areas than in rural areas, though fatality rates are
     higher in rural areas for all classes of rider and driver.

2.   In urban and in rural areas, casualty rates per vehicle-
     kilometer are much higher for riders of two-wheeled
     vehicles than for drivers of other vehicles.

3.   The severity of injuries is greater in rural areas.

4.   Since the 1967 Road Safety Act came into effect there has
     been a decline in the total number of casualties despite
     an increase in motor traffic.

5.   It is claimed that since the Clean Air Act came into force
     there has been a significant decrease in the frequency
     of inland thick fog. On thick fog days the number of
     fatal and serious injury accidents is not significantly
     affected but there is almost a 50% increase in the number
     of slight injury cases.

6.   From a study of car insurance claims confirmation has been
     obtained of the hypothesis that the number of claims
     follows a Poisson distribution* about a rate linearly
     related to age of the car.

7.   From a pilot survey of the various factors causing or
     contributing to road accidents showed that vehicle
     defects were almost exclusively confined to the braking
     systems and tyres. There is also a group of accidents
     in which loss of control could not be accounted for by
     loss of consciousness, mechanical defects or the known
     behaviour of the particular car model.

The limits on possible policy strategies to reduce the hazards
of road transport, in its present form,are largely set by: the
number and skill of drivers, the design and condition of
vehicles, the condition of the roads, and the feasibility of
replacing road transport by some less hazardous form of
transport.

AIR TRANSPORT

The hazards due to air transport have several special features;
first, as with road transport, it is mainly the traveller that
is at risk, second, the traveller has very little personal
control over a hazardous condition arising, third a hazard
generated by an aeroplane can harm people on the ground, and
fourth a hazardous condition can result from politically

---

* Poisson distribution is defined as: $P(r) = \frac{a^r}{r!} e^{-a}$

   where $P(r)$ = probability of exactly 'r' events taking place and
        'a' is the average occurrence of events in a population of
        'r' events.

motivated causes such as hijacking.

There were 2 fatal accidents on United Kingdom airlines in 1968
(11). The two accidents killed 48 passengers and seriously
injured one, also 5 crew members were killed which was slightly
below the average for the years 1964-68.* Although the number
of fatalities is small compared with the 13,222,000 passengers
carried the exposure time is also short, perhaps of the order of
an hour, as the average stage length is only 468 miles. If this
risk is taken to be a uniformly distributed risk throughout the
whole population the probability of death in a year due to an
airline crash would be of the order of 1×10^{-6}. This is a
rather optimistic figure as the number of passengers is only
about a quarter of the population so the risk to a passenger
will be nearer 2.5×10^{-5}. This still keeps the risk associated
with air transport in the range where action is expected to
reduce the hazard further. To put the present accident
probability into perspective it is important to remember that
the present accident level is a considerable improvement over
earlier years, this improvement is in part due to intensive
effort that the designers and regulatory bodies have put into
improving aircraft safety. It has been stated (12) there has
been about a 20-fold improvement in aircraft safety since 1937.
Mr. H.C. Black, Senior Assistant Chief Technical Officer of the
Air Registration Board has thrown some light on the relative
importance of the various causes of accidents (13), he has
shown that 46% of the accidents are due to crew fallibility.
22.6% of the accidents are due to material failure, and 12.6%
of the accidents are due to unknown causes. Other causes of
accidents such as operations, maintenance, weather, performance,
flying qualities and sabotage each account for less than 6% of
the total accidents, weather being the highest at 5.8% and
sabotage the lowest at 0.3%. Although more recent statistics
would probably show sabotage to be of more significance. Of the
accidents due to crew fallibility 80% were due to errors during
the approach and landing phases of flight. 9% of the crew error
accidents were due to mistakes made in navigation.

Perhaps the most worrying group of accidents, because there is
an instinctive feeling that something can be done about them,
is the group of accidents due to material failure. These are
accidents originating from component failures, faults in design,
and mistakes in assembly. The findings of the Crash
Investigation Branch, such as summarised by Newton (14), give an
indication of the way these material failure accidents arise.
Among the causes of accidents that have been found in the past
are inadequate design, some structures have been made that do
not satisfy the fail safe criteria, and vital instruments have
been fitted whose dials and pointers are so arranged that they

* In 1969 and 1970 there were no accidents, in 1971 55 people were killed
  and in 1972 112 people were killed, which suggests the 1968 fatal accident
  figures are typical of recent years.
  Ref. Annual Abstract of Statistics 1973 published HMSO.

are capable of misinterpretation. There have also been a number
of accidents due to what is known as Murphy's Law. Murphy's
Law states that "If it is mechanically possible to assemble a
vital part incorrectly, someone, some day, will assemble it in
the incorrect way." Examples of this type of error are non-
return valves fitted the wrong way round, filters fitted so that
they do not filter, and end fittings wrongly fitted to a landing
gear hydraulic jack.

With the advance of high flying supersonic aircraft there will
also be the risk that the passengers and crew may be subjected
to cosmic radiation resulting from solar flares (15). To
overcome this problem it is proposed that high flying aircraft,
such as Concorde, will be fitted with devices to give warning
of the onset of solar flares. Warned by these devices the pilot
will be able to dive to lower levels where the atmosphere will
provide an element of shielding against the radiation.

In relation to air transport the limits on possible policy
strategies to reduce hazards are somewhat similar to those for
road transport, the limits are largely represented by the skill
of the aircrew, the design and condition of the aircraft, the
effectiveness of air traffic control and feasibility of
alternative forms of transport.

FACTORIES

There are approximately 8 million people employed in 206,389
factories in Britain. The factories they work in produce a wide
diversity of goods, from food to cars, and from leather goods
to paper. The hazards associated with these factories are
equally diverse. During 1968, there were 265,861 accidents in
these factories, and 369 of the accidents were fatal, that
represents about 4.5 people killed for every 100,000 employed.*
H.M. Chief Inspector of Factories divides the causes of fatal
accidents into thirteen main groups in his Annual Report (16).
The five most important groups, from the hazard point of view,
are in descending order of importance: fires and explosions,
falls of persons, movements of non-rail transport, falls of
objects, and lastly, process machinery. It is perhaps
significant that process machinery is the least significant
cause of fatal accidents of the five mentioned.

The Factor Inspectorate have examined these fatal accidents in
some detail and have attempted to establish if some precautions
could have been taken to avoid the accidents (17). They have

_____

  * In 1972 only 261 people were killed in factory accidents, which
     suggests that taking the 1968 figure may be a shade pessimistic.
     Ref. Annual Abstract of Statistics 1973 published HMSO London.

classified the persons who could have taken precautions under the following seven headings:- management only; deceased only; fellow workmen; management and deceased jointly; management and fellow workmen. In 48% of the cases they say the management could have taken adequate precautions, and in 11.5% of the cases they say the deceased could have taken adequate precautions. There is also an eighth category which defines the events as unforeseeable, or insufficient evidence to enable an assessment to be made, and this category accounts for 23.5% of the cases. One possible interpretation of this last category is that they represent the irreducible number of accidents to be associated with this kind of activity, but this is not a completely satisfactory argument. The Inspectorate also noticed a geographical variation in the incidence of factory accidents (18) and have found that the accident rate in the North East, South Yorkshire and Wales is about twice the rate in the Midlands and Home Counties.

In addition to existing hazards new techniques are frquently introduced into factory life that may cause significant harm if used carelessly. Two examples of the techniques that come under this heading are the use of radioactive isotopes and lasers; the hazards from which are well understood, and the Inspectorate are concerned that the precautions required to control them are properly applied. In addition to the new techniques being introduced understanding is improving of the health significance of many existing processes and materials. The kinds of hazards that come under this heading are:- occupational cancer in the rubber and cable-making industries, respiratory disease in cotton workers, mesothelioma* due to asbestos, carcinoma of the nasal sinuses in woodworkers, and the toxic effects of materials such as lead, arsenic, cadmium, mercury, beryllium and phosphorous.

Possible hazard control policy strategies are limited by the action it is possible to induce management and workmen to take to avoid hazards resulting from the design, construction and operation of the factories they are concerned with. To some extent policy is limited by the availability of alternative processes and their economic acceptability.

NUCLEAR POWER REACTORS

The nature of the hazard associated with nuclear reactors is rather different from the other hazards so far considered. Although the ordinary industrial hazard found in any factory is also present in a nuclear power reactor, it is the radiation hazard peculiar to reactors that should be considered their special characteristic. To date in the United Kingdom no death has been attributed to radiation from a nuclear power reactor.

* cancers spreading over the surface of the lung.

A study by Bell (27) showed that if reactors are built to achieve the relationship between probability and release of radioactivity proposed by Farmer* then the chance of a major release would be of the same order as meteorites hitting England and Wales. This suggests that it may be possible to make and operate large and complex equipment such as nuclear reactors in a way that they do not introduce a hazard greater than the natural hazards we are all exposed to.

The radiations that are of concern are those generated by the fission process in the reactor. In simple terms when the fission process is taking place, atoms of uranium are split, new atoms are formed, known as fission products, and radiations given out. The new atoms formed may be isotopes of any of about 38 elements, some of which may be radioactive such as Iodine 131. The radiations given out may be alpha, beta or gamma rays or neutrons (19).

To put the correct emphasis on radiation hazards it must be remembered that man has always been subjected to radiation from natural sources (20). For example a person living near sea level may receive a total body dose of about 100 m rad/year, and someone living in a place like Denver, which is a mile above sea level, may receive twice that dose. Also every person contains a small, but detectable, amount of Potassium 40 and other radioactive elements inside his body (21).

The effects of radiation are usually divided into two groups, the somatic effects and the genetic effects. Somatic effects are those that are manifest in the person exposed, and genetic effects are those observed in the offspring of the exposed person. The effects produced by radiation are related to the intensity of the radiation and the length of time that the individual is exposed to it. At one end of the scale of radiation effects, large doses can cause death; for example, in statistical terms, if each member of a population received a single dose of gamma radiation of the order of 500 roentgen, at least half the population would die (22). Smaller doses would have a smaller effect; with a 50r dose, acute illness would be very rare. The types of illness that radiation may produce are leukaemia, other types of cancer and cataracts. The precise type of effect depends on how the radiation has been received as well as the size of the dose. The consequences of a dose received from radioactive material that is ingested are different from the effects on the extremities of the body, such as a foot, being subjected to external radiation.

It was mentioned above that Iodine 131 was produced as a fission product, and the Windscale incident (the only reactor accident in the United Kingdom so far) showed that it was the fission product most likely to escape in large quantities after a serious

* see Appendix II.

reactor accident that gave fission products a path to the atmosphere. Iodine 131 is an isotope, the radioactivity of which, decays to half its strength every 7 days, it is volatile, and if present in the atmosphere it is easily ingested into the body through the respiratory system. Once in the body it is preferentially taken up by the thyroid gland where it can cause thyroid cancer. Because Iodine 131 is the hazardous isotope most likely to escape in a reactor accident, it is generally taken as a measure of reactor hazard in assessing the site of a particular reactor (23).

Radiation of the reproductive organs can produce genetic effects in that the radiation may modify reproductive cells and cause mutations. At doses over 600r, the mutations produced are directly proportional to the dose; below 600r the relationship between dose and the number of mutations is being questioned and may be less than directly proportional (23). Spontaneous mutation of any particular gene is a rare event but, as there are thousands of genes in each human cell, mutations occur continually in any given population. Whether man-made radiation may induce changes to forms which never arise spontaneously is not a question that can be answered categorically, but man-made radiations have their counterparts in naturally-occurring radiations. There is some evidence from experiments with simple organisms and mice that indicates more severe effects tend to occur with disproportionate frequency after exposure to radiation compared with the frequency with which they arise spontaneously. It is not clear how far this evidence can be applied to man, and the Medical Research Council expressed the opinion that mutations produced by high doses of radiation may turn out to be, on the average, somewhat, though probably not very greatly, more harmful than those which occur spontaneously (24).

Having examined the nature of the radiation hazards and their biological effects, it is necessary to relate the radiation hazards associated with nuclear reactors to the safeguards that are engineered into them. The term "engineered" is used in its broadest sense to include all aspects of design and operation of a reactor. The principles on which the engineered safeguards are based are that protection against radiation hazards can be provided by ensuring that all radioactive material is retained in the reactor, providing shielding around the reactor to attenuate any radiation that does escape, and siting the reactor so that any radiation that does escape is separated from population by a distance that will reduce the level of radiation to such an extent that there is no unacceptable hazard. The application of these principles is generally backed up by monitoring procedures that check that operators have not received doses that could cause harm, and that no radiation has escaped outside the reactor.

The features that are built into a reactor to prevent the
escape of radiation start with placing the fuel in a can; this
can prevents fission products,the major source of radiation,
escaping. If, however, the fission products escape from the
fuel can they would still have to pass through the reactor
coolant, through the reactor pressure vessel and then through
the reactor building before they could escape from the reactor
complex and become a hazard to the general public. Each barrier
that the fission products has to pass through reduces the
probability that they would escape completely. To prevent
radiations that pass through the fuel cladding and reactor
vessel being an embarrassment to the reactor operators, shields
are erected around the reactor vessel that attenuate to an
acceptable level, any escaping radiation. Reactors are also
provided with an ingenious system of controls so coupled to
devices for sensing: temperature, pressure, coolant flow,
neutron flux and gamma flux, that if any of these indicate an
unsafe condition is developing the reactor is automatically
shut down or the appropriate action initiated.

It has already been mentioned that protection against the
hazards of radiation can be provided by having adequate
separation between the source of radiation and anyone likely to
be harmed. This safety principle has been applied in the siting
of reactors. Early reactors were very conservatively sited a
long way from major centres of population. Experience with the
first few nuclear power stations has increased the confidence of
experts in their safety and reliability to the extent that a
better compromise could be made between the other important
factors in power station siting such as: availability of
cooling water, distance from demand centres, ease of access,
freedom from flooding and the amenity value of the site. This
increased confidence in nuclear power station siting has
increased to such an extent that the CEGB is building a 1,300 MW
reactor at Seaton Carew four miles from the centre of West
Hartlepool and six miles from Middlesbrough (25)(26).

Possible alternative policy strategies are: to reduce the
hazard associated with nuclear reactors by remote siting,
greater expenditure on safety precautions, eliminate the hazard
entirely by replacing nuclear reactors by some other heat source
for electricity generation.

AIR CONTAMINATION

Although records of complaints and legislation to restrict smoke
pollution go back to the reign of Edward I positive government
action to limit the hazard from domestic fires goes back only to
the Beaver Committee, which was appointed in July 1953 after the
five day London smog of December 1952 had killed 4,000 people.
The report of the Beaver Committee (28) draws attention to the
fact that there is a clear association between contamination of
the atmosphere and the incidence of bronchitis and other
respiratory diseases. The point is also made that in countries

such as Denmark, Norway and Sweden where open coal fires are virtually unknown the male death rate due to bronchitis is of the order of 5 per 100,000, that is about a twentieth of the rate in England and Wales. Assuming half of all deaths due to respiratory diseases are in some way the result of air contamination then about 50,000 a year die due to air contamination.* This suggests the probability of death due to this cause is about 10^{-3}, which in Britain is a hazard more lethal than road transport by nearly an order of magnitude, and clearly a level at which action is required.

At the time the Beaver report was written, taking the country as a whole, nearly half of the smoke in the air came from domestic chimneys. The significance of domestic smoke varies from area to area, being more important in areas where housing predominates and is particularly dense, and of less significance in areas where there is a high concentration of smoke generating industry. The feature about the discharge of domestic smoke that accentuates its hazardous characteristics is that it is discharged at relatively low level, and consequently is particularly concentrated in the air that is being breathed. The effect is even further accentuated under inversion conditions, which prevent the smoke being dispersed and often leads to foggy conditions.

The root cause of the harmful domestic smoke is the inefficient burning in open fires of bituminous coal, with a high content of volatile matter. The problem can be overcome largely by burning smokeless solid fuels or converting all heating systems to oil, gas or electricity as the primary source of heat. The gaseous discharges from industry are more complex and any of the whole range of chemicals from lead and fluorine compounds to rare metals like gold and platinum. Bugler (29) reports that where discharges of rare metals are possible industry finds it economic to provide close to 100% efficient arrestment. This suggests that as it is technically feasible to prevent air contamination the limits on possible control policy strategies are set mainly by the cost of the alternatives.

* The death rate in Britain due to respiratory disease shows a slight tendency to increase in the years 1968-1972, but there is probably a phase lag between improvements in air quality and reduction in death rates.
  Ref. Annual Abstract of Statistics 1973 Published HMSO London

CHAPTER 3
HAZARD CONTROL POLICY

To put the discussion of policy into perspective, first, the term must be defined. Following a suggestion by Heclo (30) that, "policy could be regarded as an analytic category, the contents of which are identified by the analyst rather than by the policy maker" the following definition was adopted:- policy is taken to be a category that includes the courses of action pursued under the authority of government to achieve particular ends. In this study the ends that are of interest are the control of hazards.

Considering that the basic objectives of this study is to identify the factors that influence government policy on hazard control, it would be almost meaningless to discuss policy without first considering the policy making and implementing process. The analysis of hazard control policy is built up with this in mind in the following five steps:-

1. A model of the general policy making and implementing process is postulated. The basic characteristics of the process being described in system analysis terms, and supplemented by a description of the probable roles and interactions of the various actors.

2. The technical and institutional instruments that make up hazard control in the five cases studied are outlined, and the way they have developed is sketched.

3. From the description of the development of instruments of control the proximate policy makers and policy making centres are identified.

4. An analysis of hazard control policy is made and this leads to an evaluation of the adequacy of the model of the policy making process is postulated. The evaluation is made in terms of the efficacy of the model in describing the way hazard control policy is developed and identification of the way policy formation has been influenced in the five cases studied.

5. Finally, possible future developments in hazard control policy are suggested.

MODEL OF THE POLICY MAKING PROCESS

Outline of Model

The first step then is to build up a model of the policy making process, and the initial move in building up such a model is to

identify the characteristics of the process. The approach
adopted is to identify the major systems or sets concerned, and
then to identify the components of the sets and sub-sets that
interact to form the policy making process. In the broadest
sense the universal set that embraces all the policy making
process is the set that contains all natural systems. The three
natural systems that are relevant to this study are what have
been termed (31): human activity systems, designed abstract
systems, and designed physical systems. The human activity
system, is the system that is of major interest, as it embraces
all activities that humans participate in. Designed abstract
systems include all the knowledge and philosophical systems that
exist. The designed physical systems comprise of all
manufactured articles that make up our world.

By looking at the three natural systems it is possible to
identify the components that are linked to form an authoritative
policy making process. The human activity group includes every
grouping of humans that exist. It ranges over international,
national, and family groupings. Although the hazard policy
making process in Britain is the main interest, the governments
of other countries must be included among the groups considered,
as they influence the environment in which the British policy
making process works. Policy making in Britain is centred on
the government. This leads to identification of groups such as:
Parliament, the civil service, the electorate, and local
government. Outside the government organisation, but interacting
with it, are groups such as: industries, trade associations,
professional associations, trade unions, religious bodies,
schools, universities, research organisations, clubs, societies,
and families. This indicates the human activity groups that
have to be considered.

Moving on to the designed abstract systems, which is a range of
systems that includes all systems of knowledge that have been
devised. The state of knowledge at the time a particular
problem has to be solved has a profound influence on the
solutions that can be considered. Designed abstract systems
then can be considered as factors that influence the
environment in which policy is made.

Finally, designed physical systems include all the technological
products such as cars, aeroplanes, nuclear reactors, factories,
roads, machines, houses, air traffic control systems and road
traffic control systems, which are the concern of this study.
These systems are inanimate and so do not participate in the
policy making process, but it is their existence and operation
that sometimes give rise to demands being made for the control
of hazards associated with them.

So far the groups that have to be considered have been indicated,
but not the role of each group or the relationship of one group
with another. The next step is to identify groups associated
with policy making that have sufficient connectivity to allow
them to be considered a cohesive system. The connectivity may

have developed from an interest in a common objective. In other
words, the inputs in terms of demands, and the need for outputs
or satisfaction may have provided sufficient stimulus to cause
groups to develop connectivity. Black (32) has shown for a
simple generalised model how the inputs to a system can interact
with the sub-systems to give a particular output. This model
was for a static situation, so did not allow for changes that
may take place with time and which would require a system with
either feed-back or learning ability. Roberts (33) described
in more detail a model of the British political system which
although it was for a static situation did show the
relationship between many of the groups of the resources of
interest to this study. It showed the possible flow of opinions
and demands of the electorate through political parties and
pressure groups to government and Parliament. Also it showed
possible lines of interaction between these groups and
recognised that the possible responses of the system were
constrained to some extent by the economic system and the action
of foreign political systems. The output from the government
is described as: decisions, laws, and action. One slightly
misleading feature of the Roberts model is the way that pressure
groups and political parties have been labelled the policy
making units, and the government and Parliament labelled as the
policy implementation units. No label is ever perfect, but in
this case it would probably have been accurate to describe the
government and Parliament as the policy making and policy
implementation units.

Starting from the models of Black and Roberts, the system shown
in Fig.2 has been developed to describe the British policy
making process. The outer rectangle represents the environment
in which the policy making process operates. Three elements in
the environment that influence the policy making process it
surrounds are: economic systems, foreign political systems, and
the current state of knowledge. These three elements do not
always interact directly with the hazard control policy making
process, but there may be types of policy such as: foreign
policy and defence policy, in which they interact frequently
with the policy making process. The elements may be regarded
as constraints on the policy solutions that the proximate policy
makers may adopt. The inner rectangle contains the set of seven
major components that constitute the policy making system.
These components are: the public demands, political parties,
interest groups, Parliament, the Cabinet, the civil service and
decision implementation. The two components that need some
explanation are: the public demands, and decision implementation.
Public demands is a rather nebulus component, which is intended
to represent the inputs to the system from either the general
or specialist sections of the public. The decision
implementation component is intended to represent the output of
the system in terms of actions that have the authority of the
government behind them. The starting point for the initiation
of policy proposals can be nearly anywhere in the system. The
nature of the policy decision required dictates how many of the
components are involved. Major decisions would involve most
components of the system, while minor decisions might be settled

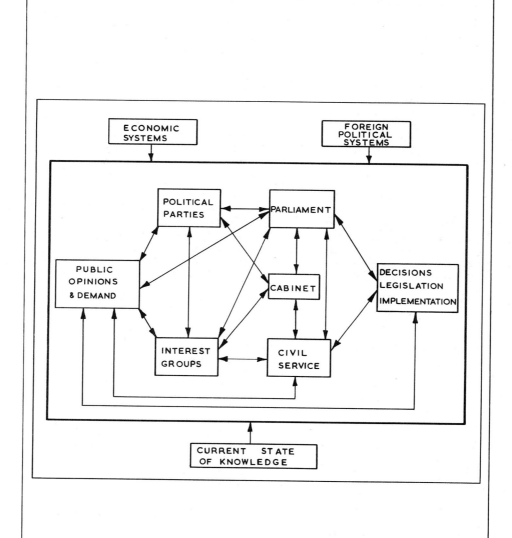

THE   POLICY   MAKING   SYSTEM

FIG. 2

departmentally in the civil service with very little involvement
of other components of the system. The Cabinet, which is taken
to include the Cabinet Office and Cabinet Committees, occupies
a node point in the system through which it is assumed all
major policy proposals pass. In passing through the Cabinet,
the proposals may be modified, at the very least the timing and
the form of action to be taken would be decided,this assumption
agrees with Walker's description of the role of the Cabinet(34).

Before straying too far into the description of the roles and
interactions of the various components of the system, attention
is drawn to the duplex nature of the links between the various
components. The links are shown to be duplex to reflect the
fact that there is some element of consultation between each of
the components in the system. In this representation no attempt
is made to show the extent of the consultation, or to show
whether the consultation is formal, such as through a committee,
or informal as through individual personal contacts.

The Nature of Demands

Having just set out the bare bones of the policy making process,
the next step is to complete the model by clothing the bones
with a description of the roles and interactions it is assumed
for each component identified. For the purpose of description
it will be convenient to start with a demand for the development
of a policy and to follow the demand through to a new policy
being adopted and implemented.

Public demands are, for this study, taken as demands from either
the general or specialist sections of the population regardless
of whether or not they can articulate their demands. It is
appreciated that the public or the part of the public that gives
rise to a particular demand may also be an integral part of one
of the other components of the system. The demands that the
public raise amount to wishes to see action taken to improve or
correct some situations. The demands do not occur spontaneously
but represent a response by the public to a situation that has
been brought to their attention in some way. A few examples of
how demands arise will illustrate how events can lead to new
demands.

Aeroplane,train, and car crashes in which many people are killed
often make headlines in the newspapers and on radio and
television both at the time of the event and at the subsequent
inquiry into the cause of the accident. The concern generated
can lead either directly or indirectly to demands for improved
safety standards. Another way in which demands may be
stimulated is by publicity given to the results of some careful
investigation that shows need for change. Typically the
investigation could be related to: the need for restricting the
use of particular chemicals in food, the need to control the
escape of certain gaseous effluents from factories, or the need
for better health checks on pilots. The contribution that the
press, radio and television make to the process is in helping
to relay demands from the public to the policy maker.

The demands may manifest themselves in the first place as questions to Members of Parliament or as topics on which interest groups express opinion and request action. The civil service will also be aware of the implications of the events and may have developed proposals for improving the situation in a way that it senses demands will require. These examples are taken intentionally from the civil domestic side of life, the demands associated with military and foreign policy matters develop in a slightly different way because discussion on these subjects is of necessity less open so there is less public involvement with the decision making process. The demands that are made will to some degree be conditioned by: the extent of the knowledge of the subject that exists, practices that are followed in other countries, and by what is considered economically practical. Although the demands will be conditioned by these factors it does not follow that the weight given to these conditioning factors by the public will be the same as that given by the policy makers.

The Role of Political Parties

Once a demand is generated it moves on through the system, it may move in parallel through political parties, Parliament, interest groups, and the civil service components of the system. Alternatively the demand may move through the system in a series fashion, one component at a time. For the purpose of this description each of these components will be considered separately, starting with the political parties.

In the context of this model the political parties are seen as one of the links between the electorate and the proximate policy makers. The features of political parties that are of interest are: the way they perform this linking function and the nature of any modification to demands that they make. The span of activities of the political parties ranges from contact at constituency level with individual voters to contact with the Prime Minister. At the constituency level the political parties will, through MP's surgeries, and ward organisations, be aware of local problems, and local views on national problems. This, in relation to hazard control problems, means that the political party will be aware of local problems such as: a particular chemical plant discharging hazardous fumes, badly designed road junctions and crossings, local concern over siting a new development such as nuclear reactors and airports. The range of problems will vary from constituency to constituency.

The constituency associations may also be used by the central party organisation to obtain opinions on policy developments that are being considered. A variation of this approach is the discussion of particular issues during Party Conferences. These discussions give the party an indication of the popular feeling on major topics, such as the care of the environment and employment conditions.

The central party organisation besides acting as a collecting
point for constituency views, also receives representations
from interest groups and incorporates these views and opinions
in advice and research for Ministers and backbench members.
This advice may be in the form of suggestions on the future
policy developments, or opinions on the acceptability of
policies that are being considered.

The role of the political parties is then to act as contact
organisation between the public and the elected parliamentary
representatives, and as an organisation that attempts to give
political weight to the opinions which it collects on particular
issues.

The Role of Interest Groups

Demands may move towards proximate policy makers by a path
leading through interest groups. Interest groups are essentially
groups of people or organisations that join together to pursue
some common objective. The groups are not necessarily
permanent organisations. They may be formed to achieve a single
objective and then be disbanded. Typical of the temporary type
of group is the kind of group formed by the residents in a
particular area to protest about the line of a new motorway.
Trade Union and Trade Associations are representative of the
permanent type of organisation. The objectives of interest
groups can be classified as economic, integrative or cultural.
Within this broad classification the objective of interest
groups may range from those with a complex mixture of objectives
to those with a simple single objective. Groups like the
Confederation of British Industry and the Trades Union Congress
could be described as having a complex range of objectives,
while a group like the National Society for Clean Air can be
considered as having a single simple interest.

One motive for people joining or forming an interest group is
to increase the force with which their views are presented. It
is an implicit characteristic of interest groups that the views
they put forward are to some extent the consensus of the views
of the individual members. Thus the demand that goes forward
from an interest group is a modified version of the demands of
the individual members. The larger type of interest group will
also have a staff organisation that will be capable of suggesting
the form that future demands could take. These suggestions after
consideration by members, may be adopted by the interest group
as demands, they would seek to see satisfied.

Permanent interest groups will have established connections with
proximate policy makers. These connections exist for two main
reasons which are: proximate policy makers regard the interest
groups as a unique source of detailed information about the
activity they are associated with, and interest groups reactions
to policy proposals are sought by proximate policy makers.
These connections allow interest groups to direct their demands
to places where at least some attention will be paid to the
demand.

In the terms of this model the role of interest groups in the
policy making process is to collect and collate the demands and
opinions of specific sections of the community, and by various
means bring these demands and opinions to the attention of
proximate policy makers in a way that influences future policy.

The Role of Parliament

The demand output from the political parties and interest groups
can move forward to the proximate policy makers in three ways:
it can move to Parliament, the Cabinet, or the civil service.
Considering first the route through Parliament, which is taken
to include both the House of Commons and the House of Lords, a
rather gross simplification of the role of Parliament is to
describe it as the body with the long term aim of improving the
common good. In practice in the short term Parliament is
concerned with satisfying those demands it sees it can satisfy.
In both cases there is a general involvement in deciding which
policies can and will be adopted. This does not mean that
Parliament approves of all policies, there is a considerable
range of policies, many of which are particularly relevant to
this study, that are conceived and implemented by civil service
departments under the authority of delegated legislation.

Demands arrive in Parliament in two main forms, which are:-
demands for modification to existing legislation that has been
found to have some unsatisfactory feature, and demands for
action to solve some newly recognised problems. The demands may
not be specified directly in these terms. They may arise
initially as questions or expressions of concern by members
about a particular situation that has arisen. These demands may
also be channeled to Parliament by the civil service and the
Cabinet as well as from political parties and interest groups.

It is the information flowing to Ministers and their Ministries
from various sources that contains some of the seeds from which
new policies grow. When this flow of information brings to a
Minister's attention some problems he considered should be
solved he will turn to his civil service staff for advice on the
possible courses of action he could follow. The demands on his
time will be such that although he will take the final decision
he will not have time to investigate the problem in depth
himself. The assessment will be prepared for the Minister
generally at about the Assistant Secretary/Principal level. If
the problem that has been uncovered is one that will require
significant legislation the Minister would probably set up a
committee of specialists to review the question. The form of
the committee depending largely on the complexity of the problem
to be handled. Straight forward changes to legislation are
likely to be considered by a departmental committee. For more
complex questions the Minister might wish to chair a committee
to consider the action required, and to have specialists from
outside his own department on the committee. For the very
complex issues, where the problem is concerned with more than
one Ministry, it is likely that the first step would be to set
up an independent committee of national authorities on the

subject to advise the Minister on the action he could take. Such a committe would be chaired by a leading national figure and would consult a wide spectrum of interested parties throughout the country. The report of such a committee would, if accepted by the Minister, form the basis for the new legislation. In some cases the department would draft legislation direct from the report, in other, more controversial cases there would first be a consultative document prepared outlining the legislation. This consultative document would be discussed with parties likely to be affected by the legislation and when all the views have been collected legislation would be drafted and submitted to Parliament. The Robens' Committee on Safety and Health at Work is an example of an independent committee of national authorities whose recommendations were discussed in the form of a consultative document before legislation was proposed.

With each type of committee there will be an involvement of the permanent civil service, although the extent and form of involvement will vary from committee to committee. At least the civil service will provide the secretarial support to the committee and advise on the procedure to be followed. The secretarial duty is quite influential as it includes drafting the final report of the committee. A greater degree of involvement of the civil service occurs in cases where Ministries have to present evidence, or make suggestions about the course of action that should be proposed. It may be that the problem has already been considered departmentally and they put forward a proposal that guides the committee to consider a particular approach.

Where the legislation required is of the kind that requires Parliament to give its approval the Minister would have to decide whether he first wanted a debate on the subject or whether he would go straight to the introduction of the required Bill. In either case he would refer the matter to the Cabinet to establish if they considered the proposals politically acceptable and the matter could be fitted into the Parliamentary programme.

The legislation would be drafted in the department and presented to Parliament either by the Minister or by a member of his Parliamentary staff to whom he delegates responsibility for guiding the Bill through Parliament. During the passage of the Bill through Parliament members will have opportunities to propose amendments. When the Bill is passed it will be the responsibility of the civil service to ensure that its provisions are implemented.

The main difference between the way Parliament deals with matters related to hazard control and other matters such as: economic, defence and foreign policy appears mainly related to the following reasons:

1.   Hazard control policy is not so important in the
      party political sense as economic, defence and
      foreign policy.

2.   Parliament is more concerned with matters related
      to public expenditure than to hazard control.

3.   Defence and foreign policy matters often require
      confidential action at the departmental levels.

This description of the role and interaction of Parliament
brings out five main features which are:

1.   Parliament provides a point of contact between
      ordinary members and Ministers that provides one
      means of bringing issues requiring policy decisions
      to the attention of Ministers.

2.   Apart from Private Members Bills the details of
      policy proposals are not worked out by Members.

3.   Parliament mainly examines and comments on policy
      proposals.

4.   Parliament has to authorise implementation of some
      types of policy proposal.

5.   The way in which Parliament deals with a particular
      problem is related to the content of the problem.

The Role of Civil Service

The two other components for demands to pass through, which were
identified in the model of the policy making system, were the
Cabinet and the civil service. Something of the role of these
two components has been suggested in the description of the role
of Parliament. The role of the civil service will be dealt with
next, leaving the role of the Cabinet the supreme policy making
body to be dealt with last.

The term civil service has been used rather loosely, as an all
embracing term, so it must be defined more carefully to prevent
misunderstanding. The civil service is intended as a generic
term * covering all the organisations that have the following
characteristics: their operations are financed by public funds,
their senior staff are appointed either by the Crown or a
Minister, their operations are carried out under the authority

* A possible alternative term is public service, but this has connotations
  of voluntary service that are not relevant to the term as it is used.
  In other parts of the text the term civil service is used in the
  conventional sense but this use is clear from the context.

of Parliament. This definition embraces in the term civil
service not only the civil service as conventionally defined,
but also local government and state owned industries such as
the Central Electricity Generating Board and British Rail.

The vital features of all the organisations that fit under this
definition is that a Minister will to some extent be responsible
for their activities and they will be organisations that have a
formal contact with a Minister. This formal contact with a
Minister will have some influence on the internal structure of
the organisation. With the knowledge that the organisation has
to be in contact with a Minister the head of an organisation
will ensure that part of the organisation maintains the
capability of providing the Minister with advice about and
information on the operations of the organisation.

This leads to identification of the organisation of units of the
civil service in a way that gives an understanding to their role.
Each unit is charged with performing some field function and is
in some way responsible to a Minister. These field functions
take many forms from providing an army, navy and airforce to
providing inspectorates such as: the Factory Inspectorate, the
Nuclear Installations Inspectorate, and the Alkali Inspectorate
that are of specific interest to this study. Providing an
inspectorate will not be the only function of a unit such as
a Ministry. The organisation will generally be much more
complex , particularly in large Ministries like the giant
Department of Trade and Industry and the Department of the
Environment. Inside these large Ministries there are several
important components. In most cases, besides the field
operation force, there will be some form of policy analysis and
review unit, a research organisation, a consultative
organisation, and a directing unit that can be identified.

As an example the Department of Trade and Industry has two field
functions relevant to this study, namely the Nuclear
Installations Inspectorate and the Civil Aviation Authority.
These two field functions show something of the spread in
administrative patterns that may be used to perform field
functions. The Nuclear Installations Inspectorate is staffed by
conventionally defined civil servants, mainly with engineering
or pure science qualifications. The Inspectorate is responsible
for licencing all nuclear power reactors, other than those owned
by the Ministry of Defence or the United Kingdom Atomic Energy
Authority. The Civil Aviation Authority is rather different.
It is an independent body responsible among other things for
the airworthiness of civil aircraft, it has some income from
fees, but its finance is backed by public funds. Senior
appointments in the Authority are made by the Minister, and he
can be questioned in Parliament about the way the Authority
performs the tasks delegated to it.

To keep up to date with technical developments in fields that
the Department is interested in it maintains several
organisations that have research as their main function, these
include: the Safety in Mines Research Establishment, the

National Engineering Laboratory, the National Physical
Laboratory, the Warren Springs Laboratory, and the United
Kingdom Atomic Energy Authority's Harwell Establishment. To
determine the most acceptable policy options the Department
operates an organisation called the Programmes Analysis Unit.
This has examined policies in many fields of interest to the
Department including in the hazard field air pollution.

Co-ordination of all the work of the Department is carried out
by the Permanent Secretary, supported by his senior staff of
Secretaries, Deputy Secretaries and a Chief Scientist.

Having set out the form of a typical department in the Civil
Service the way a policy demand flows through a department can
be postulated.

Demands can arrive in a department in several ways: they can
be passed to the department from Parliament, from the Cabinet,
or they can arise in the department as a result of its field
operations. The staff responsible in the department would then
examine the various ways of solving the problem and satisfying
the demand. In preparing these solutions the staff would,
besides drawing on their own knowledge of the subject, consult
specialists in the department and outside authorities known to
them. The task of preparing possible solutions would most
likely be the responsibility of someone at Assistant Secretary/
Principal level, and in preparing his proposal he would be
assisted by junior staff. The proposal considered most suitable
would go ahead either for approval departmentally by the Minister
or in more important cases by Parliament. The important cases
would go to the Cabinet for approval, and in lesser cases where
approving the course of action is well within the Ministers
terms of reference he would just keep the Cabinet informed of
the action he was taking.

The role of the civil service assumed in the model is simply
the conversion of demands into policy proposals, and the
implementation of the proposals that are accepted. The
interactions that the civil service develops are mainly those
with organisations likely to give specialist information, views
on the acceptability of particular policies, and those that have
to approve a particular policy.

The Role of the Cabinet

In the descriptions of the role of Parliament and the civil
service there was some indication of the part the Cabinet plays
in this model. The Cabinet is taken to consist of three
essential parts which are:- the senior executive committee of
the Government chaired by the Prime Minister, the Cabinet Office
with its staff of civil servants, and the Central Policy Review
Staff.

The first of these parts, the Cabinet as the senior executive
committee of government is perhaps most easily explained in
terms of size and time. The Cabinet generally consists of

between 10 and 20 senior Ministers. Most of these Ministers
have extensive departmental responsibilities, so can only
devote part of their time to Cabinet matters. This means that
Ministers will be concerned to make the best use of Cabinet time
by only raising what they consider are important features of
problems. The details of problems will be dealt with at the
lower levels of government. The Cabinet only meets for a few
hours at a time, so the time for discussion of any particular
problem is quite limited. The action the Cabinet takes will be
of three main forms: it will settle priorities for dealing with
particular problems, it will agree the form of action to be
taken, and it will establish responsibilities for dealing with
particular problems. This last form of action is one
manifestation of the co-ordinating role of the Cabinet.

The work of the Cabinet is supported directly by the civil
servants in the Cabinet Office. There are about 75 staff at
Principal level and over in the Cabinet Office. The staff
includes: the Chief Scientific Advisor, military specialists,
and the recently introduced Central Policy Review Staff. The
Cabinet Office also controls the Central Statistical Office, but
its activities are only indirectly related to this model. The
Cabinet Office's staff function is essentially the collection
and analysis of all the information relevant to the problems to
be considered by the Cabinet, and presenting the analysis in the
form of a paper setting out the various policy options and
implications. In preparing such papers the Cabinet Office
would call on leading experts both inside and outside the
government. For example if a decision on the type of nuclear
reactor that should be built was required the Cabinet would
expect the responsible Minister to advise them. To obtain this
advice he might form a specialist committee to advise him. He
would also expect the Chief Scientific Advisor to evaluate the
question and give an impartial view. The Chief Scientific
Advisor would consult specialists throughout the country who
were familiar and up-to-date in this particular branch of
technology. The Chief Scientific Advisor would present his
assessment to the Minister who would weigh it against the other
advice he had received before making his decision. Military
matters would be dealt with in a similar manner by the military
specialists, the main difference being that there would be
little or no discussion of the problem outside official circles.
Other subjects would be dealt with in much the same way by
appropriate officials in the Cabinet Office.

The Central Policy Review Staff is an innovation which was
introduced by Mr. Heath during 1971. The first Director General
of the Central Policy Review Staff was Lord Rothschild. At and
above Principal level there are fifteen people on the Review
Staff. Five of these hold what are known as Specialist
Appointments, that is they are not permanent members of the
civil service. Fox (35) has suggested that one function which
the Central Policy Review Staff performs is to provide the
Cabinet with independent analysis of the way policy should
develop. By independent is meant not necessarily following or

accepting uncritically the civil service or departmental line. The Review Staff have made proposals on: the way Scientific Research Councils should operate, pay and prices, Concorde, and the future of the British computer industry. This part of the Cabinet has not been in existence long enough to have made any impact on the hazard control policy considered in this study, but it may in the future make some impact on policy developments in this field.

The role of the Cabinet then is to consider the final distillate of advice on policy requirements and decide which, and in what order, the various proposals should be taken forward to implementation.

Following this description of the general policy making process the next step in the analysis is to identify the special characteristics of ᠆ne hazard control policy making process by examining the five activities, namely: road transport, air transport, factories, nuclear power reactors, and air contamination. For each activity the development of the technical and institutional instruments that have been adopted to control hazards will be indicated, and an attempt made to identify the actors concerned and their apparent goals. In each activity the identification of the proximate policy makers and policy making centres flows naturally from the examination of the technical and institutional instruments. This description of the development of policy paves the way for the evaluation of the model of the policy making process postulated and refining it to describe specifically the process for developing hazard control policy. The final section suggests possible future developments in hazard control policy.

ROAD TRANSPORT

The technical and institutional instruments that have developed to control hazards associated with road transport are particularly concerned with drivers, roads and vehicles. To help in identifying the instruments of control the essential history of the development of road transport is sketched. This is followed by a description of the institutions as they existed when the study was made, which leads to identification of proximate policy makers, policy making centres and finally to an attempt to establish what the apparent goals of the government are and have been in this field.

To put the development of road transport into perspective, it must be remembered that there has been a hundred-fold increase in the number of vehicles in use since 1910, in 1910 there were 143,877 and in 1969 there were 14,751,900 (36).

Perhaps the most restrictive instruments to control driver and vehicle were the Locomotive Acts of 1865 (The Red Flag Acts) which until their repeal in 1896, limited the speed of mechanically propelled vehicles to 2 mph in towns and 4 mph in

the country, and required the vehicle to be preceded at a
distance of not less than 60 yards by a person on foot bearing
a red flag (37). The legislation that was introduced in 1896
set the speed limit on open roads at 14 mph, or less than this
if the Local Government Board so decided (38). Over the years
governments have fairly consistently used speed limits as a
control to reduce the hazard from motor vehicles: in 1903 a
general speed limit of 20 mph (39) was imposed, in 1935 the
speed limit for built-up areas was fixed at 30 mph (40), during
the Second World War a general night-time speed limit of 20 mph
was fixed (41), and in 1967 a general limit of 70 mph was
introduced. It is tempting to speculate, although not strictly
relevant to this study, on the proposition that if the Red Flag
requirement had remained in force fewer people would have been
killed on the roads, but the country would not have made any
significant economic progress. This kind of speculation
underlines the way in which policy has to be a compromise
between the level of risk that is considered acceptable and the
social and economic goals that are aimed at. This theme of
compromise will be developed further in later sections of this
study.

In the early days of motoring there was no test of the drivers
ability to control a vehicle safely, and driving tests only
became compulsory in 1935 (42). Earlier attempts to introduce
driving tests as an instrument to reduce the hazards in motoring
had been frustrated (43). More recently there has been concern
that the drivers besides having the technical ability to control
a car safely are in the right physical condition to control a
vehicle on the road. Driving licences are issued subject to
certain conditions regarding age, health, and type of vehicle
to be driven. Examples of the physical requirements that a
driver must satisfy are that he must not be under the influence
of drugs and since the 1967 Road Safety Act if a person attempts
to drive a motor vehicle on a road or other public place when it
is proved by a laboratory test that he has over 80 milligrammes
of alcohol in 100 millitres of his blood he can be fined and/or
sentenced to imprisonment and he must be disqualified from
driving and have his licence endorsed (44). Also the driving
hours of drivers of goods and passenger vehicles are limited by
law to ensure as far as possible the driver is not too tired to
be in charge of a vehicle.

Even skilled drivers can become involved in accidents if the
roads are not adequate for the traffic using them. Over the
years road construction and maintenance policy has been subject
to a number of influences, of which economic considerations
have been particularly important. Apart from the road systems
that the Romans left there was no unified national approach to
road improvement until the beginning of the 20th century. In
the 16th and 17th century, parishioners could either work on
the roads themselves to keep them in repair or make a payment to
the parish authorities. Each year parishioners elected an
unpaid surveyor who was responsible to the Justice of the Peace
for maintaining the roads. There was a conflict of interest to
the extent that local people tended to want soft roads on which

to walk their sheep and cattle to market without risk of laming them. Travellers wanted hard, firm roads for the sake of vehicles and horses' hooves. The local Justices of the Peace tended to be sympathetic with local interests so the roads were frequently in poor condition for travellers (45). Towards the end of the 17th century, and the beginning of the 18th century, as the industrial output made greater demands on the roads, the administrative technique adopted for improving the roads was the Turnpike Trust. In their heyday, between the middle of the 18th century and the 1830s, there were some 1100 Turnpike Trusts collecting tolls on and administering about 22,000 miles of toll road. Between 1821 and 1836 Parliament did vote money directly for road improvement, this was for Telford to rebuild the London to Holyhead road which had been mismanaged by 26 Turnpike Trusts. The incentive for this measure is suggested to have been administrative and military as it was the main road to Ireland (46). The development of railways took traffic away from the roads, and consequently Turnpike Trusts went bankrupt and gradually disappeared, with the result that roads were neglected. An Act of 1862 enabled parishes to club together in highway districts to form Highway Boards, which could, if they chose, employ paid officials. In 1880 county councils were established and took over the responsibility for the main roads of the county unless urban authorities chose to continue to maintain their main roads with the aid of a county grant.

The next major development was the establishment of the Road Fund in 1909, financed by vehicle taxation, and controlled by a Road Board. The idea being that the Road Board would distribute the Fund's income to local authorities for road improvement and new road construction, maintenance being paid for out of the rates. This proved an ineffective method of improving roads and the Board was wound up in 1919 and replaced by the Ministry of Transport. At this stage a standing Statutory Committee called the Road Advisory Committee was established to advise the Minister. In the inter-war period the responsibility for initiating new roads and improvements lay with local councils, and the Ministry of Transport's part was to administer grants and to advise on standardisation of warning signs, and methods of road construction. It was only during this period that banking on corners was introduced on roads, although it had been used for many years on railways. A number of road schemes were started as a form of unemployment relief (47).

Concern about road accidents began to be taken seriously in the statistical sense, from 1922 onwards when Divisional Road Engineers of the Ministry of Transport began reporting serious accidents in their area. Where it was held that road conditions were a contributory cause of an accident steps were taken to have the highway authority effect some improvement. 1922 was also a notable year, for another reason, it was the year in which the first full scale traffic census was taken in this country.

The first major step in road development after the Second World War was the Trunk Roads Act of 1946 which brought the total

mileage of road the Minister was responsible for to 8,190 miles
(48). Other important acts were the Special Roads Act of 1949,
and the Highway Act 1959 that allowed the building of special
roads restricted to certain forms of traffic, such as are known
as motor-ways.

In 1948 the final report of the Committee on Road Safety was
published, and as a result of this report a new permanent
committee was set up under the chairmanship of the Parliamentary
Secretary of the Minister of Transport with members nominated by
other departments, local authorities, police, road users'
associations, the TUC, the National Union of Teachers, and the
Royal Society for the Prevention of Accidents. One
recommendation of the Road Safety Committee that was implemented
in 1957 was for the reduction in number of road crossings, and
for those crossings that remained to be more clearly identified.

Although there has been a considerable programme of road
improvement and motorway construction the Royal Automobile Club
(49) in commenting on the Ministry of Transport's White Paper
on "Roads for the Future - The New Inter-Urban Plan for England",
published in May 1970, found it necessary to point out that the
Ministry had itself estimated that the mileage of seriously
overloaded trunk roads would be approximately 2,250 miles by
1970.

This leads to the question of what share of the available
resources should be allocated to road works, and road safety
such as crash barriers on motorways. The methods of evaluating
the value of crash barriers used by the Ministry of Transport
were summarised in the Economist (50). Essentially the argument
was that if the value of human life is taken at £5,000 (in 1970
values) then the cost of installing the barriers is greater than
the anticipated benefit from lives saved, and the view expressed
by the Economist was that £5,000 was far too low a value for
human life. There must have been some rethinking on these
figures as crash barriers were later introduced (51).

Road improvements can be delayed by local opposition as shown
by R. Gregory's study (52) of the problems of finding an
acceptable route for the M4 from Maidenhead in Berkshire to
Tormarton in Gloucestershire. Study of the route started
seriously in the late fifties, but the road itself was only
completed in 1972. The delays resulted from the appeal
machinery built into the road approval procedure being fully
exploited by objectors. By law the Minister is bound to give
due consideration to the requirement of local and national
planning, including the requirements of agriculture. The
objectors in this case included Warmsley Rural District Council,
The Vale of White Horse Preservation Society, the Downs
Preservation Committee, the Kennet Valley Preservation Society,
Reading Council, Oxfordshire County Council, the Royal Fine Arts
Commission, the National Parks Commissioners, and the Chilterns
and South Oxfordshire Preservation Society.

The third area in which control can be exercised to reduce road transport hazard is in the design, construction and use of vehicles. Regulations have been made dealing with (53): steering gear, stability, suspension, brakes, exhaust pipes, speedometers and lighting of vehicles. In general terms the regulations are aimed at; ensuring that vehicles can be operated on normal roads without causing inconvenience to other road users, that the vehicle is stable under most conditions it will be subjected to, and that fuel is carried in a way that minimises the risks associated with such hazardous material. The braking requirement calls for two efficient braking systems, this shows some recognition of the statistical argument that duplication of a system can improve the overall reliability. The Road Traffic Act of 1960 made it unlawful to sell or supply, or to offer to sell or supply, any vehicle or trailer not complying with the Construction and Use Regulation.

In order to try and prevent vehicles being used in a hazardous condition periodic testing was also introduced under the Road Traffic Act 1960. The period between the tests depending on the age and type of vehicle. Under earlier legislation the police had the authority to inspect any vehicle they thought to be in an unsatisfactory condition. It is not universally accepted that the design, construction, maintenance, and testing standards of vehicles is satisfactory. In fact, considerable concern has been expressed in the press about these problems.

A particularly active agent for polarising concern on this subject has been the American, Mr. Ralph Nader. Mr. Nader's efforts to bring about improvements in car safety were drawn to public attention by the publication in 1965 of his book "Unsafe at any speed", which claimed that many American cars, and particularly General Motor's Corvair model were dangerous (54). General Motors went to such lengths to disprove Mr. Nader's claims that he took legal action against them. The result of this action was that General Motors paid Mr. Nader $425,000 as compensation for the invasion of his privacy (55). American cars are not the only cars that have been the subject of Mr. Nader's criticism, detailed criticism was made of some Volkswagen models (56). Mr. Nader also made a general attack on the British Government's failure to compel British car manufacturers to fit safety devices on cars exported to America (57). This attack is interesting in that it brought forth a ministerial response, and promoted discussion in Parliament. Mr. Peyton, Minister of Transport, is reported (58) as saying in reply to Mr. Nader that the government was hoping to move towards safety standards that are uniform with those of other countries, and that in Britain the death rate per million vehicle miles is lower than in the United States of America. Similar points were made by Mr. P. Walker, Secretary for the Environment, in reply to questions from Mr. Edelman and Mr. Osborn in the House of Commons on the 17th June 1971 (59).

The British vehicle test procedure has been compared unfavourably with that operated in other countries particularly Sweden. In a long article in the Sunday Times (60) a detailed comparison was

made of the Swedish annual inspection carried out by the
Bilprovning organisation, and the British Vehicle Test
performed by licensed garages. From this comparison the British
method is shown to be somewhat unsatisfactory in technical
thoroughness and availability of results.

Motoring Which is reported (61) as expressing the opinion that
the Annual Vehicle Test is a mockery, as only 44 faults out of
221 faults were found on vehicles they submitted for tests. The
Journal called for a completely new national network of
independent official testing stations. Some government
consideration was subsequently given to reducing the number of
testing stations and making purpose built test centres.
Resistance to a proposal of this type was strongly voiced by
the Motor Agents' Association (62). The resistance appears to
be based on two points, one moral and the other economic. The
moral point is that the Motor Agents do not consider they should
be in a "judge and jury" position, which they are if they both
test and repair vehicles. They also expressed the view that it
is the government's responsibility to provide test centres and
not the trade's. This is really the economic argument, they
consider the test centre should be a charge against the
government and not the motor trade. The proposal to nationalise
vehicle testing, and to improve its effectiveness still appears
to be under consideration (63), with the problem of who will
finance the new test centres being the main problem.

Looking now at the various instruments it can be seen that they
are mainly under the control of the Department of the
Environment, which now includes the Ministry of Transport. Like
all government departments, the function of the Department is
on the one hand to advise the Minister on the policy that may be
appropriate in particular circumstances, and on the other hand
to implement policy approved by Parliament. From the British
Imperial Calender and Civil Service List (64) the following
description has been built up of the civil service organisation
the Ministry had for dealing with road transport problems in
1972. The civil service side was headed by a Permanent
Secretary who had responsible to him: 2 Secretaries, 6 Deputy
Secretaries, and a Director General Highways. Below the Deputy
Secretaries were a number of Under Secretaries, one of which
was in charge of the Road Safety and Vehicle Safety Group. The
Road Safety and Vehicle Safety Group was divided into seven
Divisions which were:- Road Safety (General) Division, Road
Safety (Local) Division, Road Safety Traffic Division, Driving
and Motor Licences Division, Vehicle Safety Division, Mechanical
Engineering Division and Vehicle Inspection Division. Each of
these Divisions, with the exception of the Mechanical
Engineering Division and the Vehicle Inspection Division, was
headed by an Assistant Secretary, supported by two or three
Principals. The Mechanical Engineering Division was headed by
a Chief Mechanical Engineer and the Vehicle Inspection Division
was headed by an Assistant Chief Engineer. These last two
divisions were responsible for engineering aspects of vehicle
safety and for enforcement of statutory requirements, the

mechanical condition of public service and goods vehicles and
testings of private cars. Under the Director General Highways
there were staff responsible for: highway planning and economics
highways, The Road Research Laboratory, lands contracts and
engineering aspects of road safety. From this it appears that
within the Department of the Environment the two main centres
of interest in road transport safety were the Road Safety and
Vehicle Group and the Highway Directorate.

One role of the Ministry of Transport was to co-ordinate the
activities of the local government instruments for control of
road transport, which was divided among 823 separate highway
authorities, 382 traffic authorities, 1,190 parking authorities.
This structure of local government was clearly unsuited to the
motor age (65) and the tendency since the Second World War has
been to increase the mileage of roads that the Ministry of
Transport is directly responsible for. But Public Inquiries
under the Highways Act of 1959, Road Traffic Act 1960 and the
Transport Act 1962 are the instrument that has been built into
the system to give local objectors the opportunity to express
their views before a road policy decision is taken (66).

Other formal institutional activities that can lead eventually
to policy changes are the debates and questions in parliament
and the reports of specialised committees. An example of the
use of parliamentary question to obtain an indication of the
way policy might develop, on a particular subject, was given
above by the response of the Minister to questions about vehicle
safety that were raised after a visit of Mr. Nader to this
country.

Of the various government committees that there have been on
road transport, the departmental committee that reported in 1947
is perhaps the best example, of a committee report leading to
a policy development, its report provided the foundation for the
Road Traffic Act of 1956 (67). The important function of such
committees is that they provide a forum at which interested
parties can present their views on the way particular policies
should develop. The extent to which the views influence the
findings of a committee depends very much on the weight that the
committee puts on the evidence presented.

Instruments of a slightly different type are those operated by
the police and the courts. Both are concerned more with
enforcing the law than with policy formation. The resources of
the police are limited, so the proportion of road transport
offenders that they are able to detect and bring before the
courts only represents a sample of all offenders. Perhaps the
most that can be said is that at least the sample should act as
a salutary warning to other offenders. The courts can only deal
with the cases that are brought before them.

Outside the main stream of institutions concerned directly with
the formation of policy are two types of organisation that have
an indirect influence over policy formation. One type is
represented by the Road Research Laboratory, which is

representative of the research organisations that provide the
quantitative technical evidence on which to base policy. The
other type of organisation is the interest group, which can
both influence policy and on a voluntary basis sometimes
implement policy. Detailed discussion of this latter type of
organisation is deferred until the next chapter. However, it
is appropriate here to look at the Ministry of Transport's Road
Research Laboratory.

The importance of the work of the Laboratory was recognised in
1969 when to make it a more effective component of the Ministry
of Transport's policy making organisation a Director General
was appointed charged with this specific responsibility. He was
made responsible for four directorates: those of Scientific
Studies, Economics, Statistics, and the Road Research Laboratory
(68). In 1969 the expenditure of the Road Research Laboratory
was £2.8 million, which was 0.05% of the annual expenditure on
road transport. The non-industrial staff complement of the
Laboratory was 629.

The research of the laboratory is divided between four divisions
the Traffic Division, the Design Division, the Construction
Division and the Safety Division. Although the work of the
first three divisions have some significance for the reduction
of road transport hazards it is the Safety Division that is the
most relevant.

The Safety Division research study is grouped under four
headings, statistics, vehicle studies, driver studies, and road
layout and lighting. During 1969 the statistical studies
included: the effect of vehicle testing on the number of
defective vehicles involved in accidents, the number of accidents
in the area of central Glasgow covered by the computer-linked
area traffic control experiment, accident and casualty rates on
different classes of roads, the effect of the Road Safety Act
1967 (which introduced the breathalyser test) and the effect of
thick fog on traffic flow and accidents. The vehicle studies
include examination of the influence of vehicle design on the
incidence and nature of accidents. Driver studies have examined
driver training, driver behaviour, driver abilities and their
perception of speed and distance. Vehicle lighting and
accidents at junctions have been studied by the road layout and
lighting section.

Apart from the research in its own laboratory the Road Research
Laboratory sponsored 24 research projects at universities and
other laboratories.

In addition to the co-ordination of research by the Director
General, the research programme is reviewed by the Research
Programme Review Committee. This committee, on which only the
Ministry of Transport and the Road Research Laboratory are
represented, is chaired by a senior civil servant of the
Ministry of Transport. The terms of reference of the committee
are:

"to review the programme of road research, and its
application, and to consider its adequacy,deployment,
orientation and priorities in relation to available
manpower and money"

To advise the Director of the Road Research Laboratory on the
suitability and relevance of the research of the laboratory
there are 6 committees which bring together many of the
interested parties. For example, the Committee dealing with
Road Safety brings together the Home Office, Ministry of
Transport, Medical Research Council, and representatives of the
police, County Councils, motor manufacturing and universities.

Outside the direct government organisation are the interest
groups and a typical group in the road transport field is the
Royal Automobile Club. The RAC has two committees of interest
to this argument, its Public Policy Committee and its Working
Party on Road Safety. The membership of these committees
appears to have been selected so that it is both representative
of road transport interests and has contact with proximate
policy makers.

To illustrate the width of interest represented by the members
of the Research Programme Review Committee, and the Research
Committee on Road Safety, the membership of these committees
are detailed in Appendix 1, together with details of the RAC
Public Policy Committee, and the RAC Working Party on Road
Safety.

This review of the technical and institutional instruments
concerned with road transport hazard control suggests, perhaps
rather tritely, that those closest to the formation policy are
in the Department of the Environment either as Ministers or
civil servants. Particularly important actors being the
Permanent Secretary, the Director General Highways and the Under
Secretary responsible for the Road Safety and Vehicle Safety
Group. Slightly more removed from policy making are the
committees that are set up to enquire into particular aspects
of road transport as they bring together the views of interest
groups that have no direct responsibility for forming policy
but represent those likely to be affected by the implementation
of a particular policy. The heart of policy making for road
transport is then probably in the Department of the Environment,
but centres that influence policy, sometimes quite significant
centres, exist outside the Department particularly in the
interest groups.

The question that still remains is what is government policy on
road transport hazards. Plowden (69) suggests that up to the
1950s the Government had tried to intervene as little as
possible in the affairs of the makers and users of cars.
Plowden further suggests (70) that with regard to vehicle safety
the government saw its role as that of a night watchman. To
support his argument Plowden quotes the following statement that
the Parliamentary Secretary made during 1955. "In a democratic

country, the government cannot go further in enforcing safety
regulations than public opinion is prepared to support. We
believe that inspection of vehicles would make for a reduction
in the number of road accidents but if public opinion is not
prepared to put up with the added interference and inconvenience
it is clearly unwise for us to try to legislate ahead of public
opinion"

At an earlier period the Prime Minister had quite clearly seen
it as the duty of Ministers to control road users in such a
way as to preserve life (71).

More recently the Minister of Transport, John Peyton (72),
expressed the view that a few wild drivers were the cause of
accidents, and that the problem was to enforce the law in such
a way as to make them behave sensibly. He also expressed the
view that lighting the whole length of motorways would not be
the most profitable way of spending money.

This suggests that the government has no complete and unified
policy for control of road transport hazards, and that policy
has and continues to develop in a series of disjointed
increments, and this aspect of the formation of government
policy will be discussed at the end of the section.

AIR TRANSPORT

The history of air transport is considerably shorter than that
of road transport and the development of instruments to control
the associated hazards started mainly in the decade before the
Second World War. These instruments deal essentially with four
main aspects of air transport. These four aspects are: the
certification of aircraft and operators, the licensing of air
crew, the control of routes, and the provision of an air
traffic control system to direct the movement of aircraft.

The first British air transport legislation was the Aerial
Navigation Act of 1911, which it has been suggested was passed
because of fears that aircraft flying over large crowds (such
as those which assembled for King George V's Coronation) might
descend uncontrollably and generate death and destruction (73).
This Act gave the Home Secretary power to prohibit flying over
any area which he chose to prescribe for this purpose. This was
followed by the Aerial Navigation Act 1913 and under this Act
regulations were made which closed the skies of England to
foreigners without prior diplomatic permission. The Air
Navigation Act 1920 gave powers which enabled the requirements
for certification and licensing required by the Convention for
the Regulation of Aerial Navigation signed in Paris in 1919, to
be brought into force. This implementation of the Paris
Convention is an example of the influence of foreign political
systems on British policy. International agreement was achieved
before national legislation on the subject was passed.

Between 1922 and 1933, the importance of providing a reasonable
degree of safety for pilots and passengers became a matter of
concern, and there was doubt that the arrangements for assessing
the airworthiness of civil aircraft were adequate, this led to
a committee being appointed under Lord Gorell to review the
control of private flying and other civil aviation questions
(74). Lord Gorell had been Under Secretary of State for Air,
and his committee consisted of (75) Mr. H. Balfour, MP,
Mr. Gordon-England, Mr. Lindsay Everard, MP, Lt.Col.Moore-
Brabazon, MP, Mr. Handley Page, Mr. W. Workman and Mr.W.Burkett.
This was an extremely able committee; it included three members
of Parliament, one of whom was later Chairman of the Bristol
Aeroplane Co., and Handley Page the founder of the Handley Page
Aircraft Company.

Lord Gorell's committee report contained the plea for the
formation of a single British organisation, capable of world-
wide operation, for the purpose of registration, classification,
survey and establishment of airworthiness requirements. The
title of Air Registration Board was suggested for this new
organisation, and it was further suggested that devolution of
power to it should be such that it would be autonomous and
independent to the extent that it would be unrestricted in
deciding the recommendation it should make to the Minister about
the issue, renewal or suspension of a certificate of
airworthiness for a particular aircraft.

Under the powers delegated to the Secretary of State for Air by
the Air Navigation Act of 1936 the Air Registration Board was
established in 1973. The Air Registration Board had several
novel features, these included: that the Board should support
itself from the fees it charged, and that it should be
controlled by a council consisting of 4 representatives elected
by insurers, 4 representatives elected by constructors, 4
representatives elected by operators, 4 persons elected by the
council, having an interest in aviation, but independent of the
3 groups just mentioned, and two persons appointed by the
Secretary of State, one a pilot, and the other a layman, serving
the interests of the public. Provision of "launching costs" of
£60,000 over the first 5 years, were to be put up by the
Treasury, constructors, operators and insurers, but the Second
World War upset the provision required.

The Helmore Committee 1974-49 expressed the view that no civil
aviation certifying authority could be entirely self-supporting,
and that public funds should make good the short fall, subject
to adequate safeguards. Following from this in 1949 Lord
Packenham, the Minister, put to the ARB the simple proposition
that either they accept the Helmore view that they should be
Government financed or should become financially independent of
the Government. The ARB consulted the main users of its services
who unhesitatingly preferred to pay more and retain the self-
governance which the ARB constitution provided.

The lesson that Walter Tye, Chief Executive of the ARB, claimed
to have learned from the Helmore Committee (76) was that
governments link technical independence with financial
independence.

The next review of airworthiness arrangements, the "Jay" review,
took place in 1967, and was initiated as a result of
parliamentary concern about the crashing of two similar aircraft
within the space of a day. Besides examining the safety
performance of UK operators it examined the Board of Trade and
ARB safety procedures. The report drew attention to the fact
that Board of Trade work had been hampered by Treasury
restrictions, and that responsibilities within the Board of
Trade were fragmented. One of the members of the review team
suggested that the Board of Trade safety functions and the ARB
should be unified into one body, which would be similar in form
to the ARB organisation but should have wider responsibilities.
While the ARB saw much to commend in the concept, the Board of
Trade discouraged consideration of it.

Meanwhile, a committee under the chairmanship of Sir Ronald
Edwards had been formed to inquire into methods of licensing,
regulating competition, and the changes that may be desirable
for the development of the economy and to the service and safety
of the travelling public.

At the request of the Edwards' Committee the Royal Aeronautical
Society submitted evidence to it. The paper (77) that was
submitted contained the proposal that an all embracing Air
Transport Authority should be established. It also contained
the comment that "There is some evidence to suggest that the UK
safety record, although good, is not the best in the world".
The Edwards Committee made no criticism of the performance of
the Air Registration Board, or the operational air safety
department of the Board of Trade, but was concerned about the
interfaces between the various organisations. The committee
concluded that a more extensive amalgamation of safety functions
than was considered by the Jay review was required. A White
Paper incorporating this proposal was published, but was not
debated in Parliament at once although some steps were taken to
introduce the legislation required to establish a Civil Aviation
Authority of the form outlined. The Government elected in June
1970 wanted to review the need for any legislation being
developed by their predecessors, the conclusion of the review
was that a Civil Aviation Authority was needed, and the
legislation required for its establishment was passed during
1971.

The Civil Aviation Act 1971 (78) required that a Civil Aviation
Authority be established to encourage the development of the
British air transport industry and to take over responsibility
for the regulation of safety and navigation of aircraft. The
responsibilities of the ARB were taken over by an organisation
of the Civil Aviation Authority known as the Airworthiness
Requirements Board. The composition of the governing board is
made up of representatives from aircraft manufacturers, aircraft

operators, pilots and insurers, so it is basically the same as
the old ARB Council. The members of the new Airworthiness
Requirements Board are appointed by the Civil Aviation Authority.
The Civil Aviation Authority is appointed by the Secretary of
State. Mr. John Boyd Carpenter was appointed first chairman of
the Civil Aviation Authority and was appointed in December 1971.

The current position is then, that after several attempts all
air transport safety matters are now under unified control.
Until the Civil Aviation Authority was established the ARB had
delegated to it a wide range of airworthiness assessment
functions from approval of manufacturers to licensing of pilots
and ground engineers, and from assessing the airworthiness of
aircraft to assessing the maintenance capability of airlines.

Prior to the Civil Aviation Act 1971 being passed the
certificates of airworthiness were issued by the Board of Trade
on the recommendation of the ARB (79) and the Director of
Aviation Safety in the Board of Trade was responsible for the
overall investigation and issue of Air Operator's Certificates.
The Board of Trade was also responsible for the certification of
air crew.

Until the 1971 Act the allocation of routes and regulation of
competition between airlines in order to further the development
of British civil aviation in the public interest was the
responsibility of the Air Transport Licensing Board, ATLB was
referred to as an independent body; although members were
appointed by the Board of Trade, the Chairman was paid from
government funds, and the support staff were civil servants.
It was assumed (80) that part of this policy of allocation of
routes was intended to ensure that competition was such that
safety of operation was not impaired by financial resources of
operators being weakened by severe competition eroding profit
margins. The ATLB was quoted (81) as agreeing that it has been
given no firm policy directive to follow in deciding whether or
not a particular application should be granted, although by
setting out their reasons for particular decisions they hoped to
build up their policy in a case law fashion. ATLB have also
been quoted as saying that it was inconceivable that they would
act in a way that would allow British civil aviation to develop
against the broad current of public interest.

Important aspects of the work of ATLB were that it acted as the
controlling agent for regulating the entry of foreign airlines
and represented Britain at international negotiations on air
fares. Compared with the work of the ARB and Board of Trade the
influence of ATLB on aircraft safety was of a somewhat lower
order.

The control of air traffic from the point of view of controlling
arrivals and departures at airports and eliminating mid-air
collisions is an important safety function, and in Britain was
the responsibility of the National Air Traffic Control Service,
which was under the joint control of the Ministry of Defence and

the Board of Trade. Air traffic control is now also the
responsibility of the Civil Aviation Authority and the Ministry
of Defence. The reason for the involvement of the Ministry of
Defence is that military aircraft movements have to be
co-ordinated with the movement of civil aircraft. Some
indication of the magnitude of the problem is given by the fact
that during 1970 there were 270,000 air traffic movements at
Heathrow, and that the London Air Traffic Control Centre, which
controls all the airways and route traffic in Southern England,
handled 576,000 movements in 1970 (82).

The present and proposed developments of air traffic control
have been subject to criticism inside and outside Parliament.
Mr. Leslie Hackfield, Labour MP and at the time Opposition
Spokesman on air safety, is reported (82) as saying that
Mediator (the proposed new system for air traffic control) has
taken too long to introduce and may be partially outdated
before it becomes operational. Extensive publicity has also been
given (83)(84) to the criticism that both pilots unions and air
traffic controllers have made of the Mediator system.
Mr. Hackfield's concern about Mediator was not limited to a
single Parliamentary question, he is reported (85) as having
tabled 38 questions on the subject. The Minister of Trade is
also reported (86) as praising the merits of Mediator, from
which it can be inferred that concern about the system reached
Ministerial level.

The institutional instruments for control of air transport
hazards were in 1973 centred on the Department of Trade and
Industry, and it was the Ministry that was ultimately
responsible for the Civil Aviation Authority. The Ministry was
headed by the Secretary of State for Trade and Industry and
there were four Ministers responsible to him, the Minister of
Trade and Consumer Affairs, the Minister for Aerospace and
Shipping, the Minister for Industrial Development and the
Minister for Industry. The civil service side of the
Department was led by a Permanent Secretary, and from the
information in references 86 and 87 the following description
of the organisation he controlled has been constructed. The
first rank of officials below the Permanent Secretary consisted
of four Secretaries, one for each Minister. Below the
Secretaries were 13 Deputy Secretaries, a Chief Economic
Advisor, a Chief Scientist, the Solicitor, the Director of
Statistics and the Chief Scientific Advisor (Energy) who was
also the Chief Inspector of Nuclear Installations. (The role
of the Nuclear Installations Inspectorate is discussed in the
section dealing with Nuclear Power Reactors). Seven of the
Under Secretaries were responsible for Divisions of the Ministry
to some extent concerned with air transport. These Divisions
were: the Space Division, Air Division, Aerospace (Assessment
and Research)Division, Concorde Division, Civil Aviation
Division 1, Civil Aviation Division 2 and Civil Aviation
Division 3. There were also the Civil Aviation Safety Advisers
Division headed by the Civil Aviation Safety Advisor, and the
Accidents Investigation Branch headed by a Chief Inspector. The
Civil Aviation Division 1 to 3 were concerned mainly with

commercial and economic aspects of civil aviation, policy on air
and ground services, and the administration of state civil
aerodromes. The Division most concerned with hazard control was
the Civil Aviation Safety Advisers Division, and this Division
was concerned with: policy on aircraft operations including
supersonic and other aircraft projects, safety certification
of aircraft operators, flight crew licensing and technical
matters on aerodrome fire and rescue services. Concern with
air transport hazards within the civil service was not limited
to the Department of Trade and Industry, the Royal Aircraft
Establishments at Farnborough and Bedford together with the
National Gas Turbine Establishment came under the control of
the Ministry of Defence. Some of the work at these establishments is relevant to the development of civil aircraft.

This outline appears to confirm the view that the centre for
the highest level of policy making for air transport probably
lay in the Department of Trade and Industry, as it exercised
some measure of control over the Civil Aviation Authority.
At a slightly different level, and more concerned with the day
to day problems of hazard control in air transport, is the
section of the Civil Aviation Authority responsible for safety,
which took over the role of, and was formed from the Air
Registration Board. Walter Tye who was the Chief Executive of
the Air Registration Board became Controller of Safety in the
Civil Aviation Authority. The role that committees play in
influencing policy with regard to air transport safety
is illustrated by the succession of committees that have examined
the problem from the Gorell Committee through to the Helmore
Committee and then to the Edwards Committee (88).

The final question is, whether there is an overall government
policy on air transport safety? In technical terms the nearest
statement of general policy is the view expressed by Walter Tye
in his paper "Airworthiness and the Air Registration Board" (89)
in which he postulated that there is an acceptable level for the
frequency of accidents. He also suggested that there is an
economic optimum for expenditure on safety. In detailed failure
studies made for Concorde it is reported that an event of remote
probability was considered to be one that had a probability of
occurring of the order of 10^{-5} and 10^{-7}/hr.

Another general point is that from the time of the Helmore
Committee in 1947-49 there has been concern about the financial
independence of the ARB. The report of an interview with
Professor Keith Lucas, Chairman of the organisation within CAA
that is successor to the ARB, asserts that the ARB in its
present form is more independent than the old ARB and that it is
free from external pressures (79). This appears to imply that,
because industry supported the ARB by subscription, ARB felt
bound in some way to give undue weight to the views of industry.

Detailed policy is not fixed and permanent, but develops to
take account of new experience and knowledge. An example of
this continuous development of policy is the more frequent
checks of the condition of pilots' hearts that the Civil

Aviation Authority introduced from March 1973, as a result
of the review of medical fitness of pilots made by the
International Civil Aviation Organisation (90). The need for
some improvement in checking the health of pilots was indicated
by the fact that a contributory factor in the crash of a Trident
airliner in 1972 was that the Captain had a heart attack shortly
after take off (91).

This description of policy in the air transport field suggests:
that where possible the government prefers to have complete
control over safety regulating bodies such as the ARB, policy
is developed in a series of disjointed increments, it is
accepted that there is an optimum level of expenditure on safety
and that this level can be related to a finite accident rate,
and that foreign political systems do influence British air
transport policy. It is also possible to relate the
developments in policy to technical developments, the technical
developments taking place slightly in advance of policy
developments.

FACTORIES

For the purpose of legislation a factory is defined as any
premises where one or more persons are employed in manual labour
in any process for or incidental to: making, altering, repairing,
finishing, cleaning, or breaking up any article for the purpose
of gain or by a local authority or on behalf of the Crown, and
a place where the slaughtering of cattle is performed.

The instruments that have been developed to control factory
hazards mainly have their origin in the Factories Acts. The
factory legislation current was essentially the Factories Act of
1961, and was the product of a relatively continuous process of
revision and development of legislation to improve working
conditions that started with the Health and Morals of
Apprentices Act 1802 (92). The principle of government
inspection of factories was first introduced by Factory Act 1833,
and the implementation of the requirements of the Factories Act
was the concern of the Factory Inspectorate.

When this study was written the Secretary of State for
Employment and Productivity was the Minister responsible for the
work of the Factory Inspectorate. Each year the Chief Inspector
of Factories presented to the Secretary of State for Employment
and Productivity a report on the work of the Inspectorate.
These reports were an important vehicle for making public, the
work of the Inspectorate, and possible policy developments. In
the 1968 Report (93) the Chief Inspector mentioned that his
department was preparing for new comprehensive safety
legislation and that he had received a large volume of generally
well-informed comment on the consultative document he had issued
in December 1967. He also mentioned that there was a need to
establish an employment medical advisory service better attuned
to the needs of the future than the existing system. Attention
was also drawn to the way in which the Confederation of British

Industries and the Trades Union Congress co-operated in the
safety matters, the importance of the work of the Institution
of Industrial Safety Officers, the Royal Society for the
Prevention of Accidents, and the British Safety Council.

The Chief Inspector drew attention to his concern that there had
been a failure to recruit as many non-specialist inspectors as
were required, and that many hazards have still to be identified.

Towards the end of his letter of presentation of the Annual
Report, the Chief Inspector made the following point which
suggests he was sensitive to public opinion, as expressed through
the press, and was not satisfied with the safety of industrial
working conditions. "That such subjects (cancer resulting from
exposure to particular forms of industrial atmospheric
contamination) are of interest to the Press is, I think, an
indication of their interest and importance to the public at
large, and certainly a spur to the Inspectorate to ensure that
it does all it can to see that groups of workers in those
industries where working environments may be causing injury to
health - even though this can sometimes only be demonstrated by
statistical comparison of working population - are accorded the
safe environment to which we believe they are entitled"

The 1969 Annual Report of HM Chief Inspector of Factories (94)
throws some light on the Inspectorate's philosophy on factory
safety and on the way policy was formed. The Inspectorate's
philosophy was that the law places the primary obligation for
safety, health and welfare, with the employer, and inspection
has to be on a sampling basis. Enforcement by sampling cannot
ensure rigid compliance all the time, and the Chief Inspector's
view is that better compliance for most of the time can be
secured in most premises if the occupier is persuaded that
compliance is a matter of good practice. In some cases it was
more difficult to enforce the Factory Acts than other acts, as
offences are sometimes a matter of opinion rather than fact.
For example exceeding a speed limit in a motor car is an offence
that the police can establish with some accuracy, but an
acceptable method of operating a power press safely is something
about which there can be several opinions. The Inspectorate did
bring a number of cases to court and obtained 2,482 convictions,
and the average fine for offences concerned with health and
safety matters was £39.

The Chief Inspector expressed the opinion that the Industrial
Safety Advisory Council and the various joint advisory
committees that had been set up should in the long run help to
reduce the incidence of accidents.

To review the adequacy of the safety and health of persons in
employment, other than transport workers a committee of enquiry
was set up under the chairmanship of Lord Robens. The
establishment of this committee was announced by Mrs. Castle (95)
the Secretary of State for Employment and Productivity, during
the six hour debate on the second reading of the Employed Persons

(Health and Safety) Bill. In the part of her speech dealing
with the need for the committee Mrs. Castle expressed concern
that the law may be lagging behind changes in industry, and that
the existing machinery may not be adequate to cope with the
hazards that are associated with new technology and the
increasing scale of industrial operations. It was Mrs. Castle's
conclusion that mere consolidation and revision of existing
legislation would not be enough, and that a review was required
of the whole range of legislation from the point of view of its
effectiveness in preventing accidents and of the sort of changes
needed if a significant impact was to be made on the toll of
death, injury and ill health. Mrs. Castle also mentioned that
she intended to discuss the scope of the inquiry with the CBI
and TUC.

In due course the team to carry out the inquiry was appointed.
Lord Robens agreed to be chairman and members of the team were
as follows:-   Mr. G.H. Beeby, Ph.D, B.Sc, FRIC, Miss Mervyn
Pile MP, Mr. S.A. Robinson, Miss Anne Shaw CBE, Sir Brian
Windeyer FRCP, FRCS, and Professor J.C. Wood LLM. Lord Robens
has been: Minister of Labour, Chairman of the National Coal
Board, a union official, a Director of the Bank of England and
the Times newspaper. His interest in safety was clearly shown
in his book Human Engineering (96) in which he recognised that
legislation was not satisfactory and that the Factory
Inspectorate was often below strength. He also drew attention
to the economic advantage of improving safety by pointing out
that accidents gave rise to claims for social security payments
amounting to £96 million.

To understand the contribution that this committee of inquiry
could make to the problem of hazard control it is necessary to
consider the background of the members of the committee.
Dr. Beeby (97) was President of the Society of the Chemical
Industry in 1970, Vice President of the Royal Society for the
Prevention of Accidents in 1968 and he has held various
appointments in the rubber and chemical industry. His
publications are described as covering industrial safety,
industrial economics and business administration. Miss Mervyn
Pike is reported (97) as being a Company Director, and was
Assistant Postmaster General 1959-63, and Joint Parliamentary
Under-Secretary of State Home Office in 1964.

Mr. S.A. Robinson is reported (97) as being the retired General
President of the National Union of Boot and Shoe Operatives, a
position he held from 1947 to 1970.

Miss Anne Shaw is reported (97) as being the Chairman and
Managing Director of the Anne Shaw Organisation Ltd. since 1945.
She has also been on other government inquiries and on
government boards. Her publications are listed as "Introduction
to the theory and application of work study" (1994) and "Purpose
and Practice of Motion Study" (1952)

Sir Brian Windeyer is listed (97) as Vice Chancellor University
of London, Professor of Radiology Middlesex Hospital Medical

School, Chairman of the Radio Active Substances Advisory
Committee 1961-70, and Chairman of the Radiological Protection
Board 1970. His publications are summarised as various articles
on cancer and radiotherapy.

Professor J.C. Wood is shown (98) as being the Edward Bromley
Professor of Law at Sheffield University.

The composition of the inquiry team can fairly be described as
including representatives of: trade unions, business interests,
academic specialists in safety and law.

The committee was given very wide terms of reference (99) it was
asked to review the provision made for the safety and health of
persons (other than transport workers) in the course of their
employment. The committee concluded (100) that the present
regulatory provisions follow a style and pattern developed in an
earlier and different social and technological context. It also
concluded that the piecemeal development of regulatory provisions
had led to a haphazard mass of law which is intricate in detail,
unprogressive, often difficult to comprehend and difficult to
amend and keep up to date. The committee proposed (101) that
existing legislation should be revised and unified, and that a
National Authority should be formed which would bring together
in a single autonomus body, responsible to a Minister, all the
safety functions of the various government bodies. For the
government, Lord Gowrie stated in a House of Lords debate on
Safety and Health at Work (102) on Tuesday 30th January, that
it was still the government's intention to introduce legislation
early in the 1973-74 Session, to implement the recommendations
of the Robens' Committee. Naturally at such a stage no
indication was given of the extent to which the recommendations
would be implemented. Although earlier in answering oral
questions on industrial safety and health Mr. Maurice Macmillan
(103), Secretary of State for Employment, recognised the Robens'
Committee proposals to improve safety by meaking greater use of
workplace safety consultation, and stated this proposal was
being considered by the government. In reply to a later
question Mr. Macmillan (104) gave an assurance that it was not
the government's interpretation of Lord Robens' proposals that
there should be any weakening of statutory safety regulations.

The Robens' committee expressed concern (105) about the cost of
accidents, and expressed the view that there was need for
research into the economics of accidents and accident prevention.
The aim of this research should be to develop a more cost-
effecitve approach to the deployment of public resources for
accident prevention. Work on the effectiveness of government
intervention in occupational safety by Greenberg (106) suggests
that added investment in the Factory Inspectorate has served to
enlarge the Governmental safety establishment, but has not been
accompanied by a readily measurable decrease in the national
accident burden, except for those involving fatalities.
Greenberg very fairly points out that the statistical arguments
he used had their limitations and there was no statistical
control during his study.

In a research paper Craig Sinclair prepared for Lord Robens'
Committee he showed (107) how the cost-effectiveness approach
could be applied to safety, and that there were wide variations
in expenditure on safety in various industries. Industries that
spent most on safety were, for the cases studied, shown to
have the lowest accident rate.

In 1972 the main instrument of control of hazards in factories
was the Factory Inspectorate, and the Chief Factory Inspector
presents his Annual Report to the Secretary of State for
Employment. This is a fair indication that the Department of
Employment was the centre for the development of policy on the
control of factory hazards. The civil service side of the
Department was led by a Permanent Secretary who was supported
by three Deputy Secretaries and a Director of Occupational
Safety and Health (108). The holder of the post of Director in
1972 was, before the post of Director was created, the Deputy
Secretary responsible for the Factory Inspectorate and Health
and Welfare. It is a reasonable assumption that the Director
of Occupational Safety and Health became responsible for the
Factory Inspectorate. Below the Chief Factory Inspector there
were (109) four Deputy Chief Inspectors, seven Deputy
Superintending Inspectors and about one hundred Inspectors of
the Principal Professional Technological Officer Grade and above
In evidence to the Robens' Committee (110) it was mentioned that
there were 464 General Inspectors, although it is not clear,
this total probably includes grades below Principal Professional
Technological Officer Grade. The Inspectorate was divided into
five specialist sections: engineering, civil engineering,
medical, electrical and chemical.

A number of the Inspectors were located permanently in the
regions. Rather specialised instruments for controlling
hazards, more on a voluntary basis, were the Joint Standing
Committees that had been established for particular industries.
Professional organisations, insurance companies, trade unions
and employers were represented on these committees. The
committees were chaired by Factory Inspectors and their reports
were published as official departmental papers.

This appears to confirm the view that the factory hazards
proximate policy makers were mainly located in the Department
of Employment. In recent times the most influential committee
for polarizing opinions on future policy on the control of
factory hazards has been the Robens' Committee.

There is no single statement that embraces government policy
since the Factories Acts were first passed, in fact, as in many
fields policy has developed in a series of increments. In recent
times the Labour Government expressed concern that the existing
machinery for controlling hazards was not adequate to cope with
the new hazards that are emerging from new technology and the
increasing scale of industrial operations. This view must, to
some extent, be shared by the later Conservative Government
as they indicated that they intended to introduce legislation
to implement the recommendations of the Robens' Committee

appointed by the previous Labour Government. The statement by
the Chief Factory Inspector, quoted above, indicates official
opinion was not satisfied with the present safety arrangements,
and that some development of policy was required. The Robens'
Report suggests that the tools for evaluating the appropriate
deployment of public resources for accident prevention still
require development. So although the need to improve factory
safety has been accepted, the nature of the government's policy
to bring about improvement has not yet been made public, nor has
the extent of the resources to be devoted to such improvement
been indicated.

NUCLEAR POWER REACTORS

Of the five sources of hazard considered nuclear power reactors
are the most modern, as they have only come into being in this
country since the Second World War. It therefore follows that
the technical and institutional instruments that have been
developed to control the associated hazards are the product of
recent thinking on the form of hazard control prcedures. Before
moving on to discuss the present form of these controls a few
words must be said to put the pattern of reactor development in
this country into perspective.

The two important factors that governed the early policy for the
development of nuclear power in Britain were the defence
requirements for an independent nuclear weapon and the need to
use nuclear reactors to produce electricity to overcome the
shortages of other types of fuel. At the end of the Second
World War Britain was left without atomic bombs or reactors, and
cut off from American nuclear materials and nuclear information
by the McMahon Act (111). In order to retain Britain's position
as a major power the government gave priority to developing
resources that would enable her to have atom bomb potential (112)
To this end two major reactors were completed during 1950, and
provided the material for Britain's first atomic weapon which
was exploded in the Monte Bello Islands on 3 October 1952.
During the early post Second World War years Britain was short
of fuel and generating capacity, and it was appreciated that the
heat produced by a nuclear reactor could be used to produce
steam which could be used to drive conventional steam turbine
driven generating plant. In 1955 a White Paper (113) was
published, which proposed a ten-year programme that would give
Britain between 1500 and 2000 MWE of electricity from nuclear
reactors. The proposals of the White Paper were accepted and
work on the first two stations of the programme started in 1957.
This changing nature of the nuclear programme was reflected in
the way the activity has been controlled.

In the immediate post-war years atomic energy developments were
controlled by the Ministry of Supply, but in 1954 when it became
clear that extensive use could be made of reactors in the
generation of electricity the United Kingdom Atomic Energy
Authority was established as a public corporation (114). Two
years later the UKAEA commissioned Britain's first two power

reactors at Calder Hall, although these reactors were built
essentially to supply the plutonium that was required for
defence they did confirm the feasibility of the proposed
programme of nuclear power reactor construction. In 1958 the
Central Electricity Generating Board was established "to
develop and maintain an efficient, co-ordinated system for the
supply of electricity in bulk for all parts of England and
Wales" (115), and the CEGB subsequently became the main owner
of nuclear power reactors. In 1971 the fuel element
manufacturing parts of the UKAEA together with several reactors
were transferred to a new public corporation, British Nuclear
Fuels Limited. So the present position is that there are only
four owners of nuclear power reactors in this country, the CEGB,
South of Scotland Electricity Board, UKAEA and BNFL all four
are currently state owned organisations.

The responsibility for the safety of power reactors is divided
between the UKAEA and the Nuclear Installations Inspectorate.
The reason for this division of responsibility is due to some
extent to the way in which the nuclear industry has developed.
From the first Atomic Energy Act in 1946 the responsibility for
and control of atomic energy has rested with the Government.
Responsibility for reactor safety was more specifically
delegated to the Minister responsible by the Radioactive
Substances Act of 1948 in the following terms (116):

> "The appropriate Minister may, as respects any
> class or description of premises, or places specified
> in the regulations, being premises or places in which
> radioactive substances are manufactured, produced,
> treated, stored or used or irradiating apparatus is
> used, make such provision by regulations as appears
> to the Minister to be necessary:
>
> (a)   to prevent injury being caused by ionising
>   radiations to the health of persons employed
>   at those premises or places or other persons,
>   or
>
> (b)   to assume that any radioactive waste products
>   resulting from such manufacture, production,
>   treatment, storage or use as aforesaid are
>   disposed of safely;
>
> and the regulations may, in particular and without
> prejudice to the generality of this sub-section,
> provide for imposing requirements as to the erection
> of structural alterations of buildings or the
> carrying out of work"

In 1954 an Atomic Energy Act was passed which provided for the
establishment of the United Kingdom Atomic Energy Authority,
and modified the earlier legislation in such a way that it gave
the UKAEA responsibility for the safety of its reactors (117).
This rather unusual arrangement of one organisation being the
builder, operator and safety assessor can to some extent be
explained by the fact that the initial purpose of the UKAEA

was to produce nuclear weapons, and that practically all the experts in nuclear matters in the country were then employed by the UKAEA. The following extract from Chapter IX of the UKAEA First Annual Report 1954-56 (118) gives an indication of the policy the UKAEA adopted on reactor safety: "There are risks in atomic energy work, as there are in any scientific or industrial venture. But they are risks that are easily measureable and as well understood, as those of any other undertaking, and can be guarded against accordingly. The Authority takes the utmost care to protect their staff, and the public from any possibility of harmful effects from their operations" Both the document setting out the first programme for nuclear power and the First Annual Report of the UKAEA expressed the opinion that if nuclear power facilities are properly designed, any accidents that may occur will be no more dangerous than accidents in many other industries.

The next development in the control of reactor safety was to some extent precipitated by the Windscale incident in October 1957. This incident which occurred when an annealing process (Wigner energy release) was being performed on the graphite moderator of the core of one of the two air-cooled reactors which had been built to produce plutonium for nuclear weapons. It has been estimated (119) that 20,000 curies of Iodine 131 were released by this incident together with small quantities of other isotopes. Although technically quite a serious accident, it gave rise to claims that amounted to less than £100,000 (120) As a consequence of this incident the Prime Minister appointed two committees, both under Sir Alexander Fleck, the then Chairman of Imperial Chemical Industries. One Committee was charged with making a technical evaluation of information relating to the design and operation of the Windscale reactors, and with reviewing the factors involved in the controlled release of Wigner energy. The other Committee was asked to review the organisation within the Authority as a whole for control of health and safety and to make recommendations. This second committee was making its review at a time when it was clear that purely power producing reactors belonging to the Central Electricity Generating Board would be outside the control of the United Kingdom Atomic Energy Authority. The report of the Fleck committee (121) indicated the need for an independent body to be responsible for the safety of reactors not owned by the UKAEA. The reason for arriving at that view was that at that time the UKAEA could not claim to be disinterested on questions of siting and design since they acted as consultants to the Electricity Authorities and to industry. The Fleck report also recommended that the UKAEA should establish a national training centre to provide proper training in the techniques required by staff responsible for nuclear safety. The Fleck report recommendations were implemented in the Nuclear Installations (Licensing and Insurance) Act 1959, which besides providing for the establishment of the Nuclear Installations Inspectorate, specified that all reactors other than those belonging to the UKAEA (which were covered separately) must be insured so that all claims that might arise from the release of radioactivity material would be covered.

The position established by the Nuclear Installations
(Licensing and Insurance) Act 1959 still, at the time this book
was written,holds although subsequent legislation modified
some details. The position is that the UKAEA is responsible for
the safety of its own reactors, and that the Nuclear
Installations Inspectorate is responsible for issuing licences
to other reactor owners for siting and operation of reactors it
considers safe. The NII originally was concerned with CEGB and
SSEB reactors and the small research reactors owned by the
universities, but since the formation of the British Nuclear
Fuel Company it is also responsible for licencing their reactors.
Some indication that parliament was satisfied with these
arrangements was given by the speech that was made by Mr. David
Price during the presentation of the second reading committee
report on the 1969 Nuclear Installations Bill (122) and the
statement that Lord Sherfield made during the Committee stage
of the Atomic Energy Authority Bill in the House of Lords on
4 February 1971 (123). The relevant part of Mr. Price's speech
was as follows:- "As the Parliamentary Secretary has indicated,
the safety record of the nuclear industry has been the most
remarkable of any industry in the history of technology. I know
of no industry which has developed from scratch to the level
which it has reached today with so few casualties and so little
damage. Compared with the development of coal mining, the iron
and steel industry, the development of the motor car and the
railways, the casualties as a result of nuclear damage have
been minimal. We all know that at Harwell the danger is not
the nuclear installations; it is the Oxford to Newbury road".
The two interesting features of this statement are that it
expresses satisfaction at the nuclear safety record, and
compares the risks associated with nuclear installations
favourably with the risks in other activities.

Lord Sherfield dealt specifically with the adequacy of
administrative arrangements for ensuring the safety of nuclear
installations in the following terms: "It seems to me that the
Nuclear Installations Act 1965 sets up fully adequate machinery
for licensing, inspection and the protection of the public, and
that includes the matter of design of reactors, to which the
noble Lord particularly referred. Therefore, it is, as I think,
undesirable to set up parallel machinery, with all the
proliferation of reports, inspections, and bureaucratic
procedures, which would, if I am not mistaken, involve reporting
to a different Minister". The opinions expressed by Lord
Sheffield were endorsed by Lord Drumalbyn.

Taking reactor siting policy as an indicator of the way reactor
safety policy as a whole develops, it can be seen that the early
reactors were sited well away from major centres of population,
and that as confidence developed in the safety and reliability
of reactors they were sited closer to centres of population, in
the same way that conventional power stations are. Examples of
this newer policy of siting reactors close to centres of
population are the Heysham and Seaton Carew reactors that are

now being built. Currently the Nuclear Installations
Inspectorate assesses the suitability of sites for reactors
against the risk to the surrounding population. No engineering
plant is entirely risk free so there is no logical way of
differentiating between credible and incredible accidents.

An alternative and more general method for assessing the
suitability of a reactor for a particular site was put forward
in 1967 by Mr. Farmer, Director of the United Kingdom Atomic
Energy Authority Safety and Reliability Directorate, at the
IAEA Symposium on Containment and Siting of Nuclear Power
Reactors held in Vienna. He showed how by using probability
analysis a quantitative assessment can be made of the risks to
the population of siting a particular type of reactor in a
particular site. The Farmer criterion is applicable to all
forms of risk evaluation, and is described in Appendix II.

The details of the arguments on which reactor siting policy was
based and the incremental changes in policy have frequently
been presented at meetings such as the IAEA Symposium on
Containment and Siting of Nuclear Power Reactors and the
British Nuclear Energy Society on Safety and Siting held in
1969.

The way in which the Minister who is ultimately responsible for
reactor safety sees the advice on policy formation was stated
clearly by Mr. Marsh in answer to a parliamentary question (124)
asking whom he consulted before deciding on a change in siting
policy. Mr. Marsh was then Minister of Power and the Minister
to whom the Nuclear Installations Inspectorate reported. The
answer he gave was: "In addition to other government
departments and my own professional advisors, I consulted the
Nuclear Safety Advisory Committee. The members of this
independent committee are highly qualified and experienced men
representing both sides of industry, insurance, government
research establishments and inspectorates, the academic world,
and interests in the field of nuclear design, construction and
operation". In reply to a further parliamentary question (124),
Mr. Marsh gave the names, qualifications and business
associations of the Nuclear Safety Advisory Committee as it was
then constituted. The Chairman of the Committee was Sir Owen
Saunders, Emeritus Professor of Mechanical Engineering at
Imperial College and Past President of the Institution of
Mechanical Engineers. Details of the other members of the
committee are given in Appendix III.

Sir Owen Saunders gave the opening address to the British
Nuclear Energy Society 1969 Symposium on Safety and Siting (125)
In his address he drew particular attention to the papers
dealing with the application of probability analysis, and stated
the possibility of being able to calculate the probability
of accidents arising from numerous possible causes is a line of
approach which is well worth pursuing. The papers presented
at this symposium by the Nuclear Installations Inspectorate, by
reactor suppliers, and by the UKAEA showed that the probability
approach introduced by Mr. Farmer in 1967 had been accepted as a

useful tool for the evaluation of the safety of reactor designs
in quantitative terms. This technique has recently been
developed further by Beattie and Bell (126) who showed how for
particular conditions the costs that could be associated with
various releases could be calculated.

The institutional instruments for the control of nuclear reactor
hazards had, like air transport, their base in the Department
of Trade and Industry. The Chief Inspector of Nuclear
Installations reported to one of the four Secretaries in the
Department (87). Also the Chairman of the United Kingdom Atomic
Energy Authority, who was ultimately responsible for the safety
of the UKAEA reactors, was appointed by the Secretary of State
for Trade and Industry. There was in the Department a small
Atomic Energy Division, headed by an Under-Secretary; this
Division had a mainly co-ordinating role.

In more detail the organisation of the Nuclear Installations
Inspectorate and the Safety side of the UKAEA were as follows:-
Below the Chief Inspector of the Nuclear Installations
Inspectorate were a Deputy Inspector, three Assistant Chief
Inspectors and forty-two Inspectors of the Principal
Professional and Technology Grade and above. One of the three
Assistant Chief Inspectors was one of the secretaries of the
Nuclear Safety Advisory Committee whose co-ordinating role has
already been mentioned. In the UKAEA reactor safety was the
concern of the Safety and Reliability Directorate, whose
Director, F.R. Farmer, has already been shown to have
contributed significantly to the reactor safety argument. The
professional staff of the Directorate was about one hundred
strong. Beside being concerned with the safety of the
Authority's reactors the Directorate conducts a considerable
programme of research aimed at understanding the nature of the
safety problems that may be associated with future reactor
designs.

To summarise then the two organisations most concerned with
reactor hazard control policy were the Nuclear Installations
Inspectorate and the Safety and Reliability Directorate of the
United Kingdom Atomic Energy Authority. One body charged with
providing independent advice on nuclear safety was the Nuclear
Safety Advisory Committee. The function of this committee was
essentially to establish for the Minister the concensus of
opinion on particular nuclear safety problems. The pattern of
institutional instruments to control reactor hazards can be
traced to the recommendations of the Fleck Committee.

This leaves the question of what has been the Government's
policy on the control of nuclear reactor hazards. From the
above review of the way technical and institutional instruments
have been developed no comprehensive statement of policy emerges.
However, there are three underlying characteristics which in
the analytical sense indicate the nature of the policy that has
been pursued. First, from the beginning of the nuclear industry
in Britain there have been institutions established to control
the safe development of reactors. The form of these

institutions has only been changed when the need for change
has been apparent, for example the Nuclear Installations
Inspectorate was only established to ensure that the safety
arrangements for the reactors owned by the Generating Boards
were satisfactory when it was clear that the Generating Boards
were going to own and operate reactors. Secondly, from the
first there has been an awareness of the hazard associated with
nuclear reactors and there has been an insistence that reactors
must be subject to approval before they are operated, and
regularly inspected during their operating life. To implement
this policy the Nuclear Installations Inspectorate has been
provided with a staff that is larger per operational unit for
which it is responsible than that provided for the Factory
Inspectorate. Thirdly, from the public statements of proximate
policy makers concerned with the control of reactor safety there
are indications that the probability of risk and cost of
accidents are factors considered in deciding the acceptability
of a particular reactor on a particular site.

AIR CONTAMINATION

The instruments that had been developed to control air
contamination were mainly based on three fairly recent pieces of
legislation (127). First, and the longest established
legislation in this area was the Alkali & Works Regulation Act
of 1906, which was most recently extended by the Alkali & Works
Order 1966. This legislation was directed at the reduction of
air contamination resulting from industrial processes. Second
in chronological order was domestic smoke which was principally
subject to the controls required by the Clean Air Acts of 1956
and 1968, these in general had to be implemented by local
authorities. The third piece of legislation was the Motor
Vehicles (Construction and Use) Regulations 1969, and was aimed
at regulating the contamination from the exhausts of motor
vehicles. The regulations specify that every motor vehicle
shall be constructed so no avoidable smoke or visible vapour is
emitted.

The Alkali & Works Regulation Act required the registration and
inspection of 59 different classes of works that may generate
noxious or offensive gases. This registration and inspection
function was performed on behalf of the Minister for the
Environment by the Chief Alkali Inspector and his staff. The
annual reports of the Chief Inspector to his Minister throw
some light on the Inspector's thinking about existing and future
policy, and on the way in which the policy was implemented. In
the 1966 report (128), the first report after the Alkali & Works
Regulation Act 1906 had been extended by the Alkali & Works
Order 1966 the Chief Inspector drew attention to the fact that
several firms and trade associations had appealed to him for
delay in implementation of the inspector's requirements for
reducing air contamination. The reason for the appeals being
the difficult economic conditions facing industry at that time.
Periods of grace up to four years were allowed for existing
works to put their contamination prevention proposals into

operation. The standard required for the method of prevention
was defined as the "best practicable means". At some length
the meaning of the "best practicable means" was explained as
being the standard of treatment of emissions that is so high as
to result in little or no impact on the community and with no
scope for further major improvement. If a full and perfect
solution was not known to a particular emission problem, a
works was allowed to operate with provisional contamination
control procedures within its economic life before better
measures, which had been developed subsequently were demanded,
unless there was a justified complaint. The Chief Inspector also
accepts that delays in enforcing remedies in areas with
unsatisfactory emission problems may be unpalatable for
residents. The following quotation from the report appears to
summarise the Inspector's opinion on contamination control in
relation to financial and social implications. "The country's,
industry's and work's current financial situations have to be
weighed against the benefits for which we strive and careful
thought has to be given to decisions which would seriously
impair competitiveness in the national and international markets.
Never do we lose sight of our ultimate goal - that all
scheduled works shall operate harmlessly and inoffensively and
that this state shall be attained at the earliest possible
moment".

On the subject of the special contamination on Teeside the Chief
Inspector stated he was reluctant to ask for further large sums
of money to be spent to attempt to minimise contamination unless
he was convinced beneficial results will be obtained. To try
and find a solution to the problem the Inspectorate was
co-operating in an interdisciplinary investigation of the
problem. The problem was still so bad five years later as to
warrant a feature article in the Daily Telegraph Magazine (129)

In his 1967 report (130) the Chief Inspector mentioned that
during the year assistance was given with the preparation of a
Private Member's Clean Air Bill, and that preliminary steps
were taken with proposals for a new Alkali Order. The stage
that had been reached with the new order was consultation with
trade associations and works concerned to survey the range of
processes involved and to try and formulate suitable definitions.
The next stage envisaged was to consult with local authority
associations and other interested parties and to follow these
consultations with a Public Inquiry.

The Chief Inspector also reported that during 1967 difficulties
had again been experienced by his inspectorate in gaining
implementation of its requirements due to the prevailing
economic conditions.

The Chief Inspector found it necessary to mention that there
was an increase in the number of complaints about emissions, and
that there was an increasing tendency for these complaints to be
voiced through Members of Parliament rather than through the
Alkali Inspectorate or Public Health Departments.

In the 1967 report the Inspector also deals at some length with
his policy towards enforcement and prosecution. In his
preamble to this discussion the Chief Inspector stresses that
all members of the Inspectorate are highly qualified scientists
with a basic chemical background of a university degree or
equivalent level, and that many members also have qualifications
in chemical engineering. The general plan of campaign was for
the Chief Inspector, assisted by his two deputies, to formulate
broad national policies, after discussion with representatives
of industry, and for these policies to be applied at site level
and in detail by the inspectors and individual works
managements. In cases when difficult technical problems were
encountered,industry was offered a partnership with the
Inspectorate in finding solutions. It was stated that on only
three occasions in the past 47 years have court proceedings
been brought. The opinion was expressed that prosecution was
form of public punishment which the Inspectorate does not shrink
from applying to industry. Prosecution loses its impact if used
too frequently. Legislation and prosecution cannot of
themselves solve air pollution problems. It was claimed,
abating air pollution was a technological problem to which the
solution can be found by scientists and engineers, operating
in an atmosphere of co-operative officialdom.

In discussing the contribution of local authorities it was
stated that only the larger local authorities with specialist
staff were able to deal with other than the more straightforward
tasks of smoke abatement.

In the 1968 report (131) the Chief Inspector stated that
progress with the preliminary steps for a new Alakali Order had
been slow partly due to pressure of other work and partly
because the time was considered a little inopportune for opening
up a new field for accelerating air contamination control. The
special feature of this report was the length at which the Chief
Inspector dealt with the cost of air contamination control,
he reported the result of a survey of the capital cost and 10
year working costs of air contamination control in the scheduled
processes. The 10 year working cost was reported as £324
million or £32 million per year which was small compared with
the estimated annual cost of air contamination of £250 million
per year given in the Beaver Committee Report (132). The
Chief Inspector also expressed the view that it would be naive
to say that coal burning has such evil air pollution results
that it must be replaced by the use of other less offensive
forms of energy.

The Chief Inspector reported that a team from the United Kingdom
Atomic Energy Authority Establishment at Harwell were
collaborating with the Ministry of Technology's Warren Spring
Laboratory and local Tees-side bodies to investigate the
composition of Teeside mists with the object of providing better
information to assist decision taking on the contamination
prevention action required in the area. Numerous complains had
been registered about the smog that occured on 16 May and was
intensified by the emission caused by the breakdown of an

ammonia scrubber at a chemical plant in the area.

In his 1969 report (133) the Chief Inspector again reported that
new legislation had been further delayed. Attention was drawn
to the fact that the problems of air contamination control were
mainly economic and that if money were no object there would be
very few unresolved problems, as the technical solutions to
prevention are almost all known, and some aspects of the amount
of pollution that can be tolerated can be settled
administratively or by politically determined criteria.

The Chief Inspector mentions that a Standing Royal Commission on
the Environment was to be set up to give advice to the
Secretary of State and Parliament, and he reminds the Minister
that two important advisory bodies already exist in the air
contamination field and that their views are available to the
administration, the two bodies are the Clean Air Council and
the Interdepartmental Committee on Air Pollution Research.

When the Standing Royal Commission on the Environment was
established the Chairman appointed was Sir Eric Ashby, Master of
Clare College (details of the Membership of the Commission are
given in Appendix IV). The terms of reference of the Commission
are to advise on matters both national and international
concerning the pollution of the environment; on the adequacy of
research into this field; and the future possibilities of danger
to the environment. In the first report of the Commission (134)
they reviewed air, land and water pollution and concluded that
for the next year they should give priority to the problems of
the pollution of tidal water, estuaries and the seas around our
coast.

Attention was drawn by the Chief Inspector in his 1969 report
(133) to a major incident at a large plant for the production of
ammonia when, due to a series of accidents with plant, gas
containing sodium arsenite was released. This was thought by
the chemical company to be a purely local incident confined to
the works and the plant was restarted without reporting the
incident to any external authority, although it was later stated
that the company intended to report the incident after it had
completed its investigation. Before this was done 50 cattle on
an adjoining farm became sick and 4 cattle died of arsenic
poisoning. A local newspaper reporter heard of the incident and
found that it had not been reported to the Authorities. His
enquiries led to the incident being fully investigated by local
authorities and the Alkali Inspectorate. During the course of
this investigation it was found that vegetables in gardens three
miles away from the works were contaminated.

On the importance of pressure groups the Chief Inspector makes
the following comment "Pressures from conservationists,
anti-pollutionists and the public have helped us to raise
standards, gain better enforcements and introduce beneficial
legislation. The National Society for Clean Air has been
outstanding in this field of activity. To me our priorities are
clear and as stated in the 1965 Annual Alkali Report, the major

air pollution prevention tasks are domestic emissions, internal combustion engine exhausts, grit and dust and sulphur dioxide".

Before leaving the subject of the annual reports of the Chief Alkali Inspector there are a few general comments that should be made on the contents of the reports, these are as follows:-

1. The reports show that at the end of 1969 there were 1,691 registered works.

2. During 1969 the inspectorate made 9,563 visits and inspections.

3. During 1969, 2,018 quantitative analyses were made of gases evolved from process operations.

4. The Inspectorate investigated 412 complaints during 1969.

5. Apart from the Clean Air Council the Standing Royal Commission on the Environment and the Interdepartmental Committee on Air Pollution Research, the Inspectorate is represented on the committee on the Disposal of Solid Toxic Wastes, the Interdepartmental Committee on Major Industrial Hazards, and the National Society for Clean Air Advisory Panel.

6. The reports do not present comprehensive statistics of air pollution measurements throughout the whole country.

7. The reports make no attempt to relate deaths due to respiratory diseases to changes in levels of air pollution.

There is no exactly similar annual report to the Chief Alkali Inspectors report dealing with domestic fire smoke. However, the Clean Air Year Book (135) covers very nearly the required topics. It draws attention to the fact that during 1969 to 1970 supplies of solid smokeless fuel were critical and 16 local authorities were forced to seek suspension of smoke control orders. It was stated that Lord Robens, then Chairman of the National Coal Board, went so far as to suggest that the government should not authorise any further smoke control areas being brought into operation until adequate solid smokeless fuel was available.

The 1968 Clean Air Act extends to the smoke control provisions of the 1956 Act in a way that empowers the Minister to require a local authority to make Smoke Control Orders. The Year Book reports that there are many councils whose target date for completion of their smoke control programmes are as late as 1980 and in a few cases even later. Perhaps one reason for delaying the introduction of Smoke Control Orders is that local authorities have to find funds for financing their part of the grants that they are required to make towards the cost of adapting fireplaces to burn smokeless fuel.

There are about 120 research programmes in universities and other research establishments aimed at obtaining the scientific and technical understanding of air contamination that is required before an effective policy can be postulated for providing a completely satisfactory means of controlling air pollution (136). These research programmes can be grouped under five main headings which are: abatement of emissions of contaminants, dispersion and distribution of contaminants, effects on health, effects on animal life, vegetation and materials, and methods of measurement.

The Minister of Transport under his Motor Vehicles (Construction and Use) Regulations 1969 requires (127) "that no person shall use or cause or permit to be used on a road any motor vehicle from which any smoke visible vapour, grit, sparks, ashes, cinders, or oily substance is emitted if the emission thereof causes or is likely to cause damage to any person injury to any person who is actually at the time or who reasonably may be expected on the road, or is likely to cause danger to any such person as aforesaid". There have been proposals put forward about the standards that future motor vehicle engines will have to satisfy, and some attempts have been made to ensure that vehicles at present on the road do not emit unacceptable exhausts. It is at present too early to assess the success of these preliminary steps.

Clearly, as the Chief Alkali Inspector has recognised there has been public concern about the extent of air contamination in Britain. This concern has sometimes taken the form of open criticism of the effectiveness of the Alkali Inspectorate's control of contamination of the air. Typical of this criticism is the detailed case that Jeremy Bugler presented (137). The Robens' Committee also gave consideration to the problem of air contamination, mainly as it affected safe working conditions, and considered the role of the Alkali Inspectorate. The Committee recommended that the Alkali Inspectorate should be incorporated in the proposed unified safety authority, as this would allow better co-ordination of scientific and technical support facilities as well as more efficient deployment of inspection resources. The fact that the Committee made proposals of this nature rather suggests that they were not entirely happy with the current organisation of the Alkali Inspectorate.

The Alkali Inspectorate was in 1972 part of the Department of the Environment. The structure of this Department was briefly indicated in the section dealing with Road Transport. Below the Secretary level there were three organisations concerned with air contamination, they were: the Alkali Inspectorate, the Central Unit on Environmental Pollution, and the Directorate of Research Requirements (64). The Alkali Inspectorate was headed by the Chief Inspector,whose salary was between that of an Assistant Secretary and an Under Secretary. The Chief Inspector was supported by three Deputy Chief Inspectors and fifteen District Inspectors. According to the Robens' Report (139) the authorised establishment of Inspectorate in 1971 was 36 with an establishment of a further four in the Scottish

Industrial Pollution Inspectorate. The Inspectorate was
concerned with about 2000 establishments, so there was about one
inspector for every 50 establishments. This was only about one
sixth of the number of establishments each Factory Inspector was
responsible for. But it was significantly greater than the
number of establishments each inspector in the Nuclear
Installations Inspectorate was responsible for. The Central
Unit on Environmental Control was headed by a Chief Scientific
Officer and appears to have been responsible for co-ordinating
central government work on the control of environmental
contamination. The function of the Directorate of Research
Requirements was the analysis of research projects and
programmes including those for research in environmental
contamination, so it appears that this Directorate was concerned
particularly with the allocation of resources to contamination
research.

Outside the central government organisation, there were in
several local government authorities staff concerned with air
contamination. The staff was not uniformly distributed
throughout the country and was mainly concentrated in the larger
authorities.

From the above it can be seen that the main permanent centre of
policy making was the Department of the Environment, with the
Alkali Inspectorate playing an important role. Local
Authorities did have a role to play, but this was mainly the
local policy of implementing the Clean Air Acts in their own
area. Two committees that have been concerned to some extent
with air contamination policy, are the Standing Royal Commission
on the Environment chaired by Sir Eric Ashby, and Lord Robens'
Committee on Safety and Health at Work.

There have been a number of statements on the nature of policy
on air contamination, perhaps the most comprehensive was that
contained in the report on the Control of Pollution presented
at the United Nations Conference on the Human Environment, in
Stockholm, during June 1972 (140). The three main parts of the
policy were that:the "best practical means" were used to
prevent the escape of offensive gases, each case was judged on
its merits; and that public opinion was the deciding factor in
the extent to which any other form of pollution is controlled.

It is the vagueness of the term "best practical means" that has
been questioned. Concern being felt that this approach was
applied to the economic advantage of industry. Mr. Eldon
Griffiths, Under Secretary of State for the Environment defended
the "best practical means" approach in the following way: (141)
"We have developed a practice of collaboration between
government and industry which pays off. By applying the policy
of "best practical means" we have a dynamic policy. Technical
innovations will continue, and "best practical means" in 1975
will be much tougher than those of 1965. In other words, the
concept of "best practical means" is a dynamic one which gets
tighter year by year as technical innovation is available.

Frequently administrative standards applied by law are overtaken
by technical innovations".

Progress in the creation of smokeless zones has been slow in
the 16 years since the Clean Air Act of 1956, was passed, only
one third of the premises in the country are covered by smoke
control orders. The creation of smoke free zones was the
responsibility of local authorities, so perhaps the slow
progress was an indication of lack of public interest in the
problem. An indication of the concern in Parliament about air
contamination is given by the fact that a Bill to increase the
powers of the Alkali Inspectorate in cities was presented on
28 February 1973, and ordered to be read a second time (142).

What then in practical terms does this policy amount to? The
government provides an inspectorate to ensure that the "best
practical means" are used to minimise contamination of the air
by discharges to the atmosphere by industry. The size of the
inspectorate, in terms of inspectors per establishment to be
inspected, was smaller than the Nuclear Installations
Inspectorate but greater than the Factory Inspectorate. This
may be because those responsible for allocating manpower in the
Civil Service regarded the significance of air contamination as
less than nuclear reactor hazards but greater than factory
hazards. The acceptability of the precautions taken to
prevent contamination escaping from a particular process appears
to have been subject to negotiation between the owner of the
process and the Inspectorate. Because of the size of the
Inspectorate they cannot evaluate the contamination problems in
the same depth as the Nuclear Installations Inspectorate deal
with reactor hazard problems. In addition to providing funds
for an Inspectorate the government provided money and resources
for research into problems associated with air contamination.
The third aspect of the policy is that the government provided
funds through local authorities to assist in the creation of
smoke free zones. The creation of these zones has only
proceeded slowly, perhaps due to public apathy.

There appear to be four constraints on the controls that can be
exercised to reduce the hazards from air contamination, these
constraints are: recognition that a particular contaminant may
be a hazard, operating and engineering the process that produces
the contaminant in such a way that releases are within acceptable
limits, the financial provision necessary to implement the
the creation of smoke free zones, and the extent to which the
climate of public opinion is sensed to be in favour of a
particular purity of air.

ANALYSIS OF POLICY

The five case studies of hazard control policy just presented
are in the following analysed in four ways. First, the general
nature of the policy adopted to control hazards is examined to
establish the pattern of institutions that have been developed
for this control function. Particular attention being given to

the detail in which the controls operate, and the extent to
which the controls are either compulsory or voluntary. The
second part of the analysis is aimed at identifying the
differences in the type and range of the policy that has been
evolved for dealing with the hazards in each of the cases
considered. In this part of the analysis particular attention
will be given to the identification of reasons why the policy
should be different in different cases. The third part of the
analysis is concerned with trying to identify the ultimate
objectives that hazard control policy is aimed at. This search
for the identity of the ultimate objective is made in the hope,
which from the outset it was appreciated perhaps cannot be fully
realised, that examination of five case studies will reveal
some common policy objectives that policy makers have had in
mind when framing hazard control policy. The final part of the
analysis is directed to establishing how well the model of the
policy making system postulated at the beginning of this hazard
control policy section fits the way the five case studies
suggest the policy making system operates in practice.

The General Nature of Policy

Turning then to the first part of the analysis, that is to
examine the general nature of the hazard control policy as
revealed by the five case studies. In each of the cases
considered it can be seen that policy had developed in what
Braybrooke and Lindblom describe as a series of disjointed
increments (143), that is the policy decisions had been
adapted to the constraints of the time. Disjointed
incrementalism is the practical alternative to the synoptic
approach, which is devising policy that solves complete
problems. It may have been the policy makers intention to
devise policies that completely solved problems, but in practice
there are three main reasons why the synoptic approach fails,
these reasons are as follows:-

1.   Man's ability is limited. It is therefore very unlikely
that he ever has the knowledge or understanding to devise a
policy that will solve a social or political problem in a way
that is universally acceptable.

2.   The size of the model and the complexity of the analysis
required make it very doubtful that a completely satisfactory
theoretically devised solution can be found to more than
transiently solve social or political problems. The
simplifications that have to be made introduce errors into the
analysis that are the weakness of the solution devised.

3.   It is not possible to predict future trends with any
certainty, as so clearly argued by Karl Popper in his book the
"Poverty of Historicism", a solution may if right for today be
wrong for tomorrow (144) because the patterns of society are
always changing and developing.

The five cases considered suggest that the incremental
development of hazard control policy may have been due to the

policy maker's response to any of the following four factors:-

1.   Anticipation of public demand to restrict the
        growth of hazards associated with a particular
        branch of technology.

2.   An improved understanding of how hazards may be
        reduced.

3.   A perception of the need to gradually improve
        the quality of life.

4.   The need to remedy the defects in earlier policies.

In each of the cases considered the policy makers had
recognised that there was a need for some form of control of the
hazard. The policy for control that was adopted appears to have
the following general characteristics:-

1.   Legislation was passed.

2.   Some form of inspectorate was established charged
        with controlling the hazard.

3.   A programme of research was established aimed at
        improving understanding of the hazard and
        determining how its impact on the community could
        be reduced.

4.   Establishing some form of consultative committee
        to provide a forum where interested parties and
        those with specialist knowledge could interact
        with proximate policy makers in a way that has
        some influence on the ultimate policy adopted.

Although there were common characteristics in the policy adopted
there were differences between cases in the way the policy was
applied, for example, there was not a simple regular pattern of
inspectorates. There was a similarity between the Factory and
Alkali Inspectorates and the Nuclear Installations and Air
Transport Inspectorates but road transport inspection was a
very divided pattern rather different in character to the other
cases considered. The manufacturer has a responsibility to
market only cars that satisfy the legal safety requirements.
There are authorised garages that are responsible for ensuring
that the vehicles presented to them for testing satisfy certain
minimum requirements. The police also have a responsibility for
ensuring that vehicles are driven in a safe manner. The driver
and owner of a vehicle also have legal responsibilities to
ensure that the vehicle is only driven in a safe condition.

With nearly fifteen million vehicles on the road it would make
a considerable demand on the nation's resources to ensure that
all vehicles on the roads were in a safe condition and driven
safely. In practice it would be an impossible task to monitor
the condition of every vehicle and every driver all the time, so

the present arrangements are a compromise that leave drivers and owners a large measure of voluntary responsibility for behaving in a way that does not cause casualties. Presumably the policy makers are content that the current allocation of resources gives the optimum satisfaction of what is their perception of the public's demand for hazard control in this area.

Quite different to the form of inspection adopted for road transport was that adopted in the other four cases studied, which can be considered as two pairs each with their own special characteristics. One pair consists of the Factory Inspectorate and the Alkali Inspectorate, and the other consists of the air transport and nuclear inspectorates. The first pair of organisations were similar in the amount of attention that they gave to the hazards associated with each unit they have responsibility for, also they both appear to prefer to rely on obtaining voluntary co-operation in controlling hazards. In the case of the Alkali Inspectorate it was mentioned earlier * that the Inspectorate had been criticised for being too co-operative with industry and not effective enough in limiting atmospheric pollution.

Also when the Robens' Committee considered the role of the Alkali Inspectorate they expressed the opinion that the most satisfactory arrangement would be for the Alkali Inspectorate together with the Factory Inspectorate to become part of the proposed single national safety authority.

There is a contrast between the way that the Factory and Alkali Inspectorates operate and the way the inspectorate function was performed in relation to nuclear reactors and air transport. In the case of nuclear reactors and air transport each operational unit was subject to very detailed evaluation during design, construction and testing and was subject to re-inspection periodically throughout its life. The detailed technical involvement of these inspectorates in the design and construction to ensure requirements were satisfied was considerably more rigorous than the procedures adopted by the Factory and Alkali Inspectorate. This apparent difference between the methods of working of the Factory and Alkali Inspectorate and the Inspectorate responsible for air transport and nuclear reactors may reflect the different outlook of government on the role it should play in controlling hazards in the periods when these controls were developed.

The origins of the controls employed by the Factory Inspectorate and the Alkali Inspectorate were in the later part of the 19th century and the first decade of this century, a period that could perhaps be fairly described as one that did not welcome government interference with industry. Also, it was perhaps a period when there was little demand recognised by the policy

* see Page 61

makers for the control of hazards. The legislation only really
made an impact on new works, so general changes were slow as it
took years for works, existing at the time legislation was
introduced, to be phased out naturally as they became obsolete.
Even now when approval is sought for new works significant
weight appears to be given to economic arguments against the
elimination of all potentially hazardous conditions.

On the other hand both the nuclear and aircraft industries
started with an appreciation of the hazards involved and in a
climate of public opinion that demanded from the beginning that
positive control of hazards was exercised.

There are then at least four major reasons why the voluntary
component in hazard control policies varied from case to case.
These reasons are:

1.   Perhaps most important is the number of operational
units that have to be controlled. Where there are many units
to be controlled such as in road transport, it is not practical
to check that controls are applied universally, consequently
whatever the original policy maker's intent was, in practice it
depends on voluntary compliance for success. When a smaller
number of units is involved such as in the case of nuclear
reactors and air transport it is feasible to confirm that, to a
very large extent, each unit complies with the requirements of
the hazard control policy, so the voluntary component in hazard
control policy is smaller.

2.   The extent of the voluntary component is related to the
stage of development of the activity when the hazard control
policy is applied. The voluntary compliance component appears
to be large when the controls are applied to an established
activity, such as was the case particularly in relation to
factories and air contamination. The voluntary component is
smaller when the control is applied at an early stage in the
development of an activity, in other words before methods of
working and opinions have become rather rigidly fixed. This
smaller voluntary component due to the controls being applied
at an early stage in the development of an activity appears to
be confirmed in the cases of air transport and nuclear reactors.

3.   Variations in the proportion of the voluntary compliance
component is also related to the extent to which it is left to
other organisations to implement a particular control policy.
An example of this type of variation is shown by the way the
Clean Air Acts have been implemented. In this case it was left
to local government to implement the Acts, and because
implementation was to an extent voluntary, and there have been
variations in the interpretation of policy, this has resulted in
wide regional variations in the extent to which smoke free zones
have been adopted.

4.   Variation in the extent to which voluntary compliance is
adopted is also related to the existence of strong interest
groups, for example implementation of the Alkali and Works

Regulations Acts appear to have been modified on occasions to
suit the economic wishes of industry. This suggests that a
powerful interest group is able, by negotiation, to increase
the extent to which compliance with legislation is voluntary.

Differences in Policy

Moving now to the second part of the analysis, namely trying to
identify why the type and range of controls employed in various
cases were different, and why the policy adopted should have
allowed these variations. Some suggestion of the nature of
these variations was hinted at in the first part of this
analysis, where it was indicated that in the control of nuclear
reactors and air transport a more detailed technical assessment
was made of the hazard problems than in the other cases
considered. The extent of this technical concern was perhaps
most clearly demonstrated by the extent to which quantitative
methods of assessment are used in these two activities. In
both cases detailed quantitative analysis of the behaviour under
fault conditions was required, and it was incumbent on those
responsible for introducing a particular reactor or aircraft to
show by detailed technical argument that the requirements were
satisfied. In the case of air contamination and factories the
form of control was less detailed. The practice of presenting
very detailed technical justification for a particular activity
was not followed, nor was there staff in the inspectorates to
deal adequately with such technical justifications for all the
operational units they were responsible for. So control in
these cases was exercised by setting out the requirements that
the owners or those responsible for the particular activity
must satisfy. These requirements were often specified in
general qualitative terms, although in some cases such as the
discharge of known hazardous substances the ways in which they
may be released were carefully specified. The size of the
inspectorates was such that inspections, to confirm whether or
not the specified conditions were being maintained, may have
been less than annually. So the inspectorates may not have
known the current position in a particular operational unit.
In the case of smoke from domestic fires, even where there was a
smoke control order, the local authority to whome responsibility
for control was delegated may not have sufficient staff to
perform the inspection function adequately.

The characteristics of the hazard control policy in relation to
road transport were rather between those exhibited by policy for
nuclear reactors and air transport and those exhibited by
factory and air contamination control policy. Vehicles and
drivers had to satisfy certain specified conditions. Vehicles
were subject in many cases to annual inspections, but drivers
only had to pass a driving test once in a lifetime. Because of
the size of the problem a very great deal of responsibility for
ensuring that vehicle and driver comply with all safety
requirements was left with the owner and operator. This means
that although the requirements may be comprehensive and cover
everything from size and number of lights to condition of tyres,
and from the health of the driver to the effectiveness of the

of the braking system there may be periods when the requirements
are not satisfied.

This part of the analysis suggests conclusions that are very
similar to those indicated by the first part cf the analysis.
Certainly the four factors that were suggested as having some
bearing on the voluntary component in hazard control policy,
also have some influence on the detail in which hazards are
controlled in the five cases studied. So these four factors
will now be examined in relation to the type and detail of the
control policy adopted:

1.   The first factor was the number of operational units to
be controlled. When the number was small, as in the cases of
nuclear reactors and air transport, the responsible
inspectorates were able to examine the proposal in fine detail
and were able to ensure at frequent intervals that the
requirements were being satisfied. In the case of larger
populations of units such as factories and the various units
that give rise to air contamination, the requirements were not
specified in such detail, nor were the inspectorates large
enough to check frequently that all units satisfy the
requirements. In the case of road transport, there was a very
large population of operational units, and although there were
very detailed and complex sets of requirements to be satisfied
due to the size of the enforcement problem the compromise had
been accepted that compliance with the requirements depended to
a large extent on the operator.

2.   The second factor was the stage of development of the
activity when the control was first applied. In the cases of
air transport and nuclear reactors, controls on their operation
have been exercised from the beginning and detailed government
control has always been a feature of their development. On
the other hand, the controls on factories and air contamination
were only applied at a later stage in their development,
consequently it has taken time to correct some of the inherited
doubtful practices. Also there has been resistance, on economic
grounds, to imposing detailed controls rigorously and quickly.
Road transport, after its very early days, was left for a
considerable period with a great deal of freedom in the way it
developed. This freedom was perhaps, as was suggested in the
work of Plowden referred to earlier, a matter of political
philosophy at the time. The more recently imposed hazard
control requirements have been difficult to apply continuously
more because of the size of the population to be dealt with than
for any other reason.

3.   The third factor was the extent of delegated
responsibility for control. In three of the cases considered,
namely; air transport, nuclear reactors and factories, the
controls were very directly and fully under central government
control. In the first two cases the control was exercised in
fine detail and was constantly supervised. However, in the
case of factories the control was somewhat weaker, but this was
being reviewed. In the case of road transport and air

contamination a considerable degree of responsibility was
delegated to organisations outside the central government, and
in these cases it appears that the control over the hazards was
considerably weaker.

4.   The fourth factor influencing variation in range and
detail of control was the existence of strong interest groups.
Certainly the Chief Alkali Inspector in his reports recognised
that the policy he wished to apply was from time to time
modified in response to representations he received from industry
The airline pilots were also very concerned about the way hazard
control policies related to air transport were implemented and
made their views known. In each of the cases studied some kind
of forum was identified in which the various interested parties
could interact with proximate policy makers in a way which
influenced the formation of new policies or modified existing
policies. It is therefore reasonable to assume that interest
groups do influence policy in detail and this aspect of policy
making will be examined in detail in the next chapter.

An additional factor, that is perhaps relevant to the hazard
control policy, as applied to the activities considered, is the
extent of the technical appreciation of the way hazards can be
controlled in the particular activities. In the air transport
and nuclear reactor fields and to a lesser extent in the road
transport field, there appears to be sufficient understanding
of the nature of the problems to allow quantitative methods
to be used to evaluate the significance of the hazards and to
indicate the most efficacious ways of reducing hazards. In the
other activities studied there was not the same general
application of these methods, although Lord Robens in reviewing
safety at work did draw attention to the need for greater study
of quantitative methods in the safety fields.

Objectives of Policy

This leads to the third part of the analysis, which is an
attempt to identify the ultimate objectives of hazard control
policy. It is a little easier first to say what hazard control
policy is not, rather than specifying exactly what it is.
Certainly hazard control policy is not aimed at complete
elimination of all hazards, in the absolute sense, such a goal
would be impractical, as the demand such a policy would make on
resources would prejudice the satisfaction of other important
demands. On the other hand, activities are not now allowed
to develop freely in ways which create hazards greater than
already exist. Policy appears to have been aimed at somewhere
between these two extremes, and could be described as being
aimed at reducing the burden of hazards on the public. The
precise level of hazard that is considered acceptable is to
some extent the policy makers interpretation of what public
demands are or should be: these demands may not have been
openly expressed. With time opinions change and, generally,
there appears to have been an attempt to continually reduce the
level of hazards. In cases such as air transport it is possible
to see quantitatively how greater attention to safe design and

operation has brought down the accident rate over the years.

To have a hazard control policy does depend first on recognising
that particular circumstances can give rise to a hazard. So an
active hazard control policy must include an information
collecting system for sensing the development of a hazardous
condition or any change to existing hazardous conditions in the
community. In practical terms this information collecting
system can be seen to exist as a matter of policy and to have
led to various control institutions being established.

To establish the nature and concern of the public about hazards
it appears to be a matter of policy to establish some form of
consultative machinery. This machinery may take the form of
formal committees or informal contact between proximate policy
makers and interested and informed parties.

It also appears to be policy to establish a central government
specialist, in the Fulton (145) sense, inspectorate to control
the hazard in any particular activity. The size of the
Inspectorate for a particular activity appears to be related to
the significance a particular hazard is considered to have.
These specialist inspectorates appear to have a considerable
degree of autonomy in settling the control standards they expect
the activity to satisfy. In cases where the full significance
of a particular hazard is not known or a method of reducing the
hazard known, it appears to be a part of the policy for the
government to sponsor research to solve these problems.

The ultimate objectives of hazard control policy as it appears
to have been exhibited in the five cases considered, seems to
consist of five elements which can be summarised as follows:-

1.   To identify, as soon as possible, any hazards or
changes to hazards that arise in the community.

2.   For any hazard identified establish a legal framework
for control institutions to be established to keep the burden
of the hazard within limits considered to be acceptable to the
public.

3.   To ensure that adequate consultative machinery is
maintained to allow interested and informed parties to make
their views on hazard control policies known to the proximate
policy makers.

4.   To sponsor the research necessary to establish the
significance of hazards and to find methods of reducing the
hazards.

5.   To ensure that expenditure on hazard control is in
harmony with other demands for resources.

These five objectives of policy suggest a generally flexible
approach that can be adapted to particular activities and
changing demands. Only in the cases of air transport and

nuclear reactors was there a significant attempt to give safety
policy quantitative expression: although from arguments given
in an earlier section it is clearly feasible to express the
objectives of hazard control policy in quantitative terms, in
all the cases considered.

Adequacy of the Model

Having analysed the general nature of hazard control policy the
final part of the analysis is to examine the adequacy of the
postulated model of the policy making system. The five cases
studied indicate the characteristics of the hazard policy
making system and the strengths and weaknesses of the model.

The essential characteristics of the hazard control policy
making process suggested by this study are:-

1.   The direct concern of political parties with hazard
control issues is small.

2.   Parliamentary scrutiny of hazard control policy is
limited, in practical terms, mainly to examination of new
legislation which establishes or changes the way control of a
particular hazard will be organised.

3.   Cabinet involvement with the formation of hazard
control policy is low.

4.   Hazard control policy making is dominated by the civil
service, with the specialists (in the Fulton sense)* in the
various inspectorates having extensive control over the form
and application of the technical aspects of policy.

5.   Many of the demands for new, and changes to existing,
policy arise within the departments with responsibility for
hazard control.

6.   When major developments in policy are being considered
extensive use is made of independent specialist committees to
consult interested parties and advise on the policy required.
It is through these committees that interest groups develop
their influence on policy formation.

Now we can look at the model to see how well it represented
these characteristics. The strength of the model was that it
specified the operational environment in which the policy making
process works as being conditioned by: economic systems, foreign
political systems, and the current state of knowledge. The five
cases considered contained examples of how these factors
influenced the policy decisions. The reports of the Chief
Alkali Inspector, and the statements by Walter Tye, Controller

*The Fulton report did not deal specifically with the technical policy
 making role of the specialised inspectorates.

of Safety in the Civil Aviation Authority indicated the type of
influence that economic systems and knowledge had on hazard
control policy. Also the statement quoted by the Minister of
Transport in which he mentioned the government was hoping to
move towards safety standards unfirom with other countries shows
the influence of foreign political systems on British hazard
control policy.

In the model it was assumed that policy demands were refined as
they passed through the various components. The way this
refining process takes place is particularly clearly shown by
the way committees such as the Edwards Committee on Air
Transport and the Robens' Committee on Safety and Health at Work
collected the views of a wide spectrum of opinion and distilled
these divergent views into a fairly coherent policy proposal.

The study did show in qualitative terms the role and interactions of the main components in the policy making process
were much as specified in the model. The study also showed
that hazard control policy making is a specialist central civil
service sub-system of the general policy making process.

The model only considers the main components and not all the
sub-systems that make up each component. From the case studies
it is seen that the components of particular importance to
hazard control policy making are: specialist committees,
economic interest groups, technological activity, the Cabinet,
and the specialist inspectorates. The way these components
interact with each other and with the specialist civil service
to make the hazard control policy making sub-system is shown
diagrammatically in Fig.3.

This leads to consideration of the weaknesses of the model, it
is as a predictive tool that the model is weakest. The model
does not predict exactly the way each component will develop a
particular policy proposal or precisely the extent to which a
particular component will get involved with a particular proposal
Nor does the model indicate the nature of the interactions that
will take place between components in a particular case but
merely indicates the interactions that may take place. In other
words the model does not predict the working of the policy
making process with the exactness that models can predict the
behaviour of mechanical systems.

The conclusions that are suggested by this examination of the
model of the policy making process postulated are as follows:-

1.   The model is a useful guide to the analyst as it
indicates the actors and interactions that should be considered
in the analysis of the policy making process in general.

2.   The model does not indicate the components likely to be
most involved in the making of hazard control policy, in a
particular case.

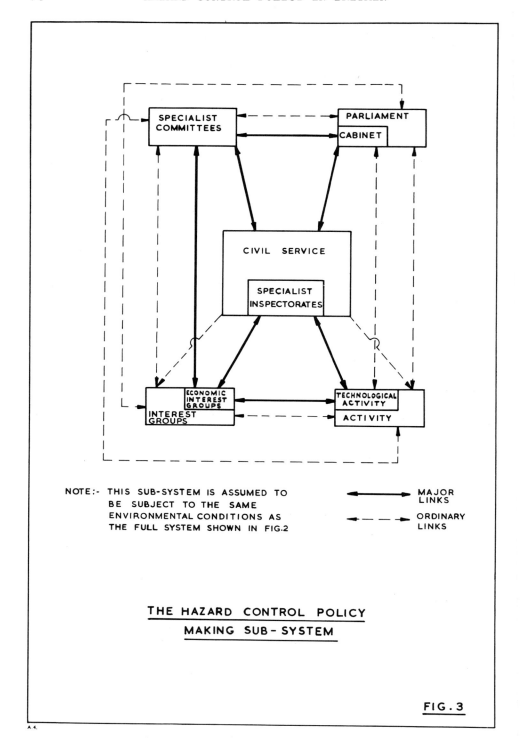

NOTE:- THIS SUB-SYSTEM IS ASSUMED TO
      BE SUBJECT TO THE SAME
      ENVIRONMENTAL CONDITIONS AS
      THE FULL SYSTEM SHOWN IN FIG.2

MAJOR LINKS

ORDINARY LINKS

THE HAZARD CONTROL POLICY
MAKING SUB-SYSTEM

FIG.3

3.   The model does not indicate in advance the nature of
the policy that will develop to deal with a particular
situation.

FUTURE HAZARD CONTROL POLICY

Having examined the nature of hazard control policy and the way
it has developed over the years; it is sensible to complete the
picture by speculating on the way policy may develop in the
future. Clearly there is no accurate way of predicting the
future, as there is no way of knowing exactly what the state of
technology and civilization will be in the years to come.
Taking the limit of this look into the future as the end of the
century, it is possible to detect a number of trends that may
continue to develop over the intervening years.

These trends are of two types: those that will give rise to new
demands for hazard control policy, and the type of policy that
may be adopted. It is the second kind of trend that is of most
interest to this study, but the first type of trend is important
because it is only when a demand is established that a policy
is developed. Typical of the demand forming trends are: the
increasing complexity of technological processes and products,
improved understanding of the means of determining the
signficance of hazards, and the increasing wish of the public to
reduce the general level of hazards. The trend towards more
unified and stricter control of hazards are typical of the
policy makers response to new demands for policy.

Before considering the possible developments in policy it is
worth looking at the demand forming trends in a little more
detail to appreciate the significance of the impact they may
have on policy developments. The demand for economic growth
leads to demands for higher productivity. Also our present way
of life means that some natural resources of minerals and oils
are being used at rates that will lead to their being exhausted
early in the next century, unless steps are taken to reduce
consumption significantly during the remainder of this century.
Both the demand for higher productivity and for alternative
materials will lead to new processes, products, and more
productive processes being sought and adopted. The new
processes will tend to be more complex, and higher productivity
versions of existing processes tend to be larger plants. New
processes and products may bring with them new hazards, and the
large production units may increase the size of existing hazards.

Research continues to reveal the hazard significance of
materials with increasing precision. Methods for predicting
the safety of processes are under continual development, and are
being more widely applied. For example the quantitative methods
for predicting system reliability that were originally developed
for the various space programmes have gradually been applied in
the aircraft, motor car and nuclear industries. These techniques
are also beginning to be applied to the chemical industry.

It seems likely that these quantitative methods of hazard evaluation will be further refined, and gradually applied over wider and wider fields of activity.

Public concern over hazards seems to be increasing, there is concern that all products sold to the public should as far as practical be hazard free. For example drugs should not have any unacceptable side effects. Food additives should not be harmful, to this end several additives have been banned in recent years. Also people are demanding safer and safer cars, to this end better brakes and tyres are now being made and seat belts are now fitted. At the other end of the spectrum unions are concerned to obtain safer working conditions for their members. Similarly industrialists will not wish to have accidents that incur costs through lost production or that damage the public image of their organisation.

In this climate of desire for less hazards, what is likely to happen to hazard control policy? Already there are signs that the advantages of unifying the control over hazards are being recognised. For example the Robens' report on Safety and Health at Work proposed a considerable measure of unification of the institutions for the control of hazards. The report recommended that: the Factor Inspectorate, the Alkali Inspectorate, and the Nuclear Installations Inspectorate should be brought together into one specialist safety (or hazard control) organisation outside the central government machine. Even if this type of unification does not come about immediately it is reasonable to forecast that unification will come about in the long term. The problems of hazard control in various fields have so many common features that it is sensible to bring all the experts together in one organisation, so that a common approach to these problems is made and uniform standards are adopted.

Forming unified organisations outside the central government machine to deal with particular functions is a form of "hiving off" that Fulton (146) suggested should be considered. In discussing the reorganisation of the Civil Service, Clarke(147) has drawn attention to the fact that a precedent for "hiving off" major regulatory functions has been set by the formation of the Civil Aviation Authority.

Currently in the aircraft and nuclear reactor field careful technical evaluation of the acceptability of the safety implications of all designs is made before they are put into operation. This careful evaluation of plants and products before they are put into use is likely to be extended. Certainly the full evaluation of chemical plants is likely to become a requirement before the end of the century. A "type approval" for road vehicles on the lines of that adopted for aircraft is a possible development in the next decade. Similar approval procedures are likely to be adopted in other fields, there is already an embryonic scheme of this nature for drugs.

A development of a slightly different type that it is possible
to foresee is that employers will have to maintain precise
records of the conditions that all employees work under, and
that each employee would be entitled to a copy of the record
of his working conditions for each employer he had worked for.
At present if a man, who has worked in several hazardous
industries, develops a disease that can be attributed to the
conditions he has worked in, it is difficult for the man to
prove the extent of his exposure and where liability may lie.
The working conditions record would overcome this problem.
This hazard record would have two other advantages: it would
make the employer think carefully about the liability he may be
building up for himself by exposing his employees to hazards,
and it would make the population in general more aware of the
nature of industrial hzards.

With the increasing incidence of acts of deliberate sabotage
that could cause otherwise safe operations to become hazardous
more attention will have to be given to protection against such
acts. The action that might have to be taken includes searching
all people entering factories and making all plant containing
hazardous material capable of withstanding sabotage.

The possible developments in hazard control are then:
unification of organisations responsible for hazard control,
detailed evaluation of plants and products to establish before
they are put into use if their hazard characteristics are
acceptable, all workers to have a detailed record of the
hazards they have been exposed to in the course of their
employment and hazardous plants to be made sabotage proof.

CHAPTER 4
ROLE OF INTEREST GROUPS

In the previous chapter interest groups were identified as one
of the actors in the hazard policy making system. In this
chapter the role of the interest groups and the way they are
organised to perform their role is examined in greater detail.
The analysis is built up in three stages. First, the
definition and classification of interest groups and their role,
as identified in relevant literature, is briefly reviewed.
Secondly, the role and organisation of a sample of interest
groups that were considered to be concerned with hazard control
policy in the five fields studied is examined. This examination
is based mainly on data that the interest groups supplied in
response to a questionnaire they were sent, supplemented in
some cases by the evidence they submitted to the Robens'
Committee on Safety and Health at Work. Finally, an attempt is
made to define in general terms the role that the interest
groups appear to play in the formation of hazard control policy.
This final section also raises some questions on the adequacy
of methods used for classifying interest groups in relation to
the role they may be expected to play.

DEFINITION AND CLASSIFICATION OF INTEREST GROUPS

Definition of Interest Groups

We can start with the definition of the term interest group
that has been used to describe groups that aim to influence
policy. Wootton (148) after looking at a number of names such
as lobby and pressure group that could be, and have been, used
to describe groups that aim to influence policy, concluded that
the terms interest group was the most suitable term. His
argument against the use of the terms lobby and pressure group
to name the groups to be studied is that usage in America had
made them into somewhat derogatory terms. Against this, it can
be argued that the term interest group is rather too passive a
term to describe the dedicated approach that some groups show
in their attempts to influence policy.

The term 'pressure group' is widely used in standard political
text books, such as Madgwick (149) and Rose (150). Although
Madgwick does draw attention to the opinion that 'pressure' may
be regarded as a loaded word, and that it implies a continuous
activity.

In his book "Anonymous Empire" (151), Professor Finer uses the
term lobby to describe the groups that attempt to influence
policy. He defends the use of the term lobby on the grounds
that the use of the word 'pressure' would imply that some kind
of sanction will be applied if a demand is refused, and that

pressure would be exerted all the time.   'Interest Group' he considers to be a too neutral and narrow term.

Moodie and Studdert-Kennedy (152) defend the use of the term 'pressure group' against the term 'interest group' on the basis that there is a risk of confusion with other established usages. Although they accepted that it is useful to be able to talk of the commercial, landed, church, business or labour interest, in the sense of an identifiable and important section of the community whose responses and welfare no government dare ignore, without necessarily meaning that it has a spokesman or organisation to represent it.   They also object to Finer's use of the term 'lobby' for much the same reason that Wootton does.

Roberts (153) equates the term 'pressure groups' with the term 'lobby' and with the term 'interest group' but suggests they all have acquired unsavoury overtones (154).   However, later he suggests (154) the terms have slightly different meanings.   A 'lobby', according to Roberts' definition, is organised and operates solely for purposes of political influence on a particular matter.   A pressure group he suggests has political functions alongisde its other functions.   Interests are the common factors which may link various individuals (or organisations) together into interest groups, which may then from time to time find it necessary to involve themselves in political activity as a lobby or a pressure group.   Taken in this sense the term interest group is endorsed as being the class name which includes lobbies and pressure groups as sub-divisions.   Defined in this way the term recognises that with time the precise role of an interest group can vary.

Finally, in this outline of the terms that have been used to describe groups that attempt to influence policy, mention must be made of Olson's (155) use of the term 'group'.   Olson defines 'group' as an all embracing term covering any group of people that share a common interest in addition to their individual interest.

Clearly there is some ambiguity in the way terms like lobby, pressure group and interest group have been used, and to prevent confusion about the way the term interest group is used in this study the following definition has been adopted:-   interest groups are the organisations that people form to pursue interests that they have in common with other people.

This definition does not in any way specify the goals that the interest groups pursue nor does it give an indication of the nature of the organisation of the group.   The next step is then to look at ways interest groups may be classified in terms of their characteristics.

Classification of Interest Groups

Wootton (156) proposed a system classifying interest groups, which was a development of that proposed by other workers in the field.   The Wootton system identifies three types of interest group, and three levels at which each group operates.

The three types of interest group identified are economic,
integrative and cultural. Under the economic heading are
considered those groups particularly associated with the
business life of the community, it includes those associated
with companies, factories and the production of goods.
Integrative groups are groups associated with the binding
together of those who perform particular roles in society. This
in itself is not a particularly neat or precise definition, but
its meaning does become clearer when it is considered in
relation to the other two groups, and a few examples are given
of the type of group that it is considered fit under this heading.
Integrative groups are not groups motivated by purely economic
or commercial interests as such groups would come under the
economic heading. Further, they are not groups associated solely
with families, religion, or education as these would come under
the heading of cultural groups. Examples of groups that fit
under the integrative heading are institutions that demand
certain standards of behaviour within a profession such as: the
British Medical Association, the Institution of Civil Engineers,
and the Inns of Court. What has been said already may indicate
the nature of cultural groups, but adding that they are
essentially groups concerned with lasting moral values makes
their definition a little more precise.

Wootton identifies three levels of interest group activity,
these levels are called: first-order, second-order and third-
order. The first-order level of activity is the small local
group, such as a single company or a local interest group
dealing with a single specialised subject of purely local
importance. The second-order level covers major interest group
working at regional and national level, but excludes the very
highest level of interest groups that are included in the
third-order groups. The third-order groups work at the highest
level, and represent their members at national and international levels. In their dealings with government third-order
groups deal at Ministerial level and Permanent Secretary level.

Moodie and Studdert-Kennedy (157) recognise that pressure groups
(as they call interest groups) have been classified in various
ways. In an attempt to classify groups more clearly they suggest
classifying groups as either promotional or formal role. Their
argument being that it is important to stress the object or
activity of the group in the label, rather than an interest
attitude distinction. Promotional groups are those formed to
put forward a common view on a specific issue, and formal role
groups are made up of members who initially at least serve
some common role. This classification does not appear, in the
form presented, to be a sufficiently detailed method of
classification as it only identifies two of the differences that
exist between interest groups.

Hanson and Walles (158) use the terms defensive and promotional
for classifying groups. Defensive groups being defined as those
groups like trade unions, which may be concerned to protect the
interests of their members. Promotional groups being defined as

those that seek to advance a cause, in which their members are
interested. The promotional type of group also has the
characteristic that they are concerned with problems that have
little to offer in terms of personal gain to those who support
it. This classification is rather coarse and does not take very
far the identification of the many different types of interest
group that make up the whole spectrum of interest groups. To
refine this crude classification, a little, Hanson and Walles do
suggest the use of the terms partial and exclusive, proposed by
Duverger (159). Partial being used to denote the group that
performs some service for members, other than their purely
defensive or promotional activity. Exclusive groups being those
that are only concerned with political activity to achieve their
goals. This suggestion of partial or exclusive categories for
groups is a possible way of differentiating between multi-
function and single function groups.

Duverger (159) refined his classification even further by
suggesting the structure of a group can be classified either as
mass groups or cadre groups. Mass groups being organisations
such as workers' trade unions and ex-servicemens' organisations.
Cadre groups being groups such as learned societies that appeal
to small but influential social categories.

Roberts (160) suggests a method of classification similar to
but more complicated than the Wootton method. Besides
dividing groups into lobbies, pressure-groups and interest
groups, the important suggestion he makes is for groups to be
classified as to whether they are permanently established groups
or temporary groups. Roberts then suggests groups should be
classified according to whether they are promotional or
defensive. Additionally groups should also be graded according
to the extent they are concerned with the personal benefit of
their members or with the general benefit. Three other
classifications that Roberts suggests can be used are to divide
groups according to whether they have mainly economic, social-
cultural or political interests. The economic and social-
cultural classifications are fairly straightforward, but the
political classification needs some explanation. The way
political interest is defined in this context is to mean any
group concerned mainly with altering the political process.

The classification of groups that Olson suggests (161) is rather
different in character to those that have already been discussed.
Three types of classification were postulated, they were:
privileged groups, intermediate groups and latent groups. A
privileged group is defined as one in which some of the members
have an incentive to see a collective good provided in the sense
that they will derive some benefit from the good. An
intermediate group is one in which no single member will derive
so much benefit from the end being sought that he would have the
incentive to provide it for himself. The latent group is
essentially a large group that needs a special incentive before
its latent power is realised or mobilised. The advantages of
the Olson classification appear to be more related to the

analysis of the economic role of groups and the latent nature
of some groups.

For the purpose of this study it was decided to adopt the
Wootton method of classifying groups as it was a fairly
comprehensive method. Although it was recognised that it might
be possible to refine the method by taking account of the
significance of the permanence of the group. This question of
the adequacy of the method of classifying groups is returned to
at the end of this section.

The Literature Of Interest Groups Related to Hazard
Control Policy

Now moving on to consider the role of interest groups as
described in the literature. The first point to note is that,
apart from road transport and air contamination, there is very
little discussion in the literature of the role of interest
groups and their influence on hazard control policy as
considered in this study. In view of this the role of interest
groups is examined in general terms.

Beer (162) stresses that interest groups' advice and judgement
is often sought by government on subjects in which the groups
have specialist knowledge. This allows the groups to influence
the detail of policies adopted. A particular case of this
nature that is quoted is the advice that the Minister of
Transport seeks from the Society of Motor Manufacturers and
Traders in relation to the revision of regulations on the
construction and use of motor vehicles. It is also recognised
that some interest groups have influence in the field of
economic policy, and, that interest groups that can employ
economic sanctions are in a fairly strong bargaining position.*
Beer (162) also suggests that from time to time governments need
the active co-operation of interest groups, particularly for the
enforcement of voluntary controls. Beer was, in fact, talking
mainly about economic controls, but similar enforcement problems
do arise in the hazard control field.

Beer considers (162) recognition of the role of interest groups
by government amounts to approval of their role and encouragement
of their activities. This view leads to consideration of the
channels for communication that the government establishes to
help interest groups interact with proximate policy makers.
These lines of communication often, in practical terms, take the
form of departmental advisory committees. An example of these
committees that was mentioned by Beer was the National Advisory
Council for the Motor Manufacturing Industry. Recognition of
groups by asking them to attend formal committees may enhance

---

*In the later sections trade associations and trade unions are classified as
  economic interest groups, but in practical terms sanctions are an asset
  of the trade unions rather than the trade association.

the group's reputation and help to establish their role as a
body to be consulted.

Stewart (163) shows how the need to consult interest groups is
an established part of the procedure adopted in forming
delegated legislation. Several cases are cited, one example,
particularly relevant to this study, being the way the Ministry
of Civil Aviation first discussed a new regulation with
interested bodies such as the pilots' Association, aircraft
operators and trade unions.* The point is also made that in
order to prevent the prerogative of Parliament being infringed,
Parliament tends to limit consultation of a proposal to the
period before legislation.

Stewart (164) also shows how some groups try and obtain
representation in Parliament, in order to bring their views more
directly to the attention of policy makers. The case is quoted
of a Secretary of the British Medical Association who resigned
to become a Member of Parliament. The implication being that on
election he would still owe some allegiance to the organisation
he had been associated with. Stewart also (164) shows how by
approaches to Parliamentary candidates before election, interest
groups attempt to obtain some form of commitment that the
candidate will support their interests.

In his conclusions about the role of interest groups Stewart
suggests (165) "Pressure groups (interest groups in the terms
of this study) are necessary to the government of our complex
society. The coherent expression of opinion they render is
vital. They have become a fifth estate, the means by which
many individuals contribute to politics. Without them
discontent would grow and knowledge be lost. It is important
that the system of government be such that their role can be
carried out with responsibility".

Roberts (166) draws attention to the activity of interest groups
in relation to such matters as: the introduction of the 70 mph
speed limit, breathalyser tests for drivers suspected of being
intoxicated, and the siting of a major airport at Stanstead. In
examining what an interest group can achieve Roberts (167)
recognised that an important constraint is not only the amount
of money available to their own organisation but also the money
available to the government. So to some extent the success of
an interest group in a particular area depends on the success it
has in competing for scarce resources. Two other constraints
that are seen to limit the success of an interest group are:
the extent to which the group can establish satisfactory
communication with all concerned with implementing the required
changes, and the extent to which the group is able to take
advantage of the opportunities that occur for groups to present
their case in a way that will have some impact on policy.

* Consultation before regulations are made is a requirement written
  into some legislation.

The skill that an interest group needs to take advantage of situations is to some extent given more point by the following quotation from Machiavelli's "The Prince", which Roberts used to illustrate his argument, "One change always leaves the path prepared for the introduction of another".

Consideration of this maxim of Machiavelli's in relation to interest groups seems to lead naturally to contemplation of the role of interest groups in more philosophical terms and to consideration of Olson's "by-product" and "Special Interest" theories in particular. Olson argues (168) that the policy influencing role of a large economic interest group is supported mainly because they provide some other service to their members, in other words, it is a by-product of their other activities. From this it seems possible to infer, intuitively, that in performing one role an interest group could open up ways of it performing another role. It could be that because a trade union or trade association has developed contacts with the government through being involved in some discussion on, say, wages, the same organisation's views could be sought on some new safety problem in the industries they are associated with.

Olson (169) also argues that the small specialised economic groups, particularly the small trade association, appear, in the USA, to be very successful in achieving benefits such as favourable tariffs, special tax rulings and generous regulatory policies. The success of these "Special Interest" groups is attributed to the groups being unique sources of knowledge and understanding of the activity they represent. This "Special Interest" concept is in many ways just another way of expressing the opinion of Beer, mentioned above, that it is the specialist knowledge and judgement of a group that a government seeks and which establishes the group's role in the policy making system.

From this brief review it can be seen that the role of interest groups is, from the point of view of their members, to present members' views and demands. From the point of view of the government the important roles of the groups are: as a source of specialist knowledge and judgement, and sometimes as a body that can help enforce voluntary controls. The role of an interest group only becomes established in relation to the policy making system if its activities are approved and encouraged by government. Two important constraints on interest group activity that often exist are: that consultation is generally limited to consideration of proposals before legislation; and the extent that an interest group's demands can be satisfied are limited by consideration of the availability of relevant resources. Interest groups are considered to be a means of allowing individuals to contribute to politics.

ROLE AND ORGANISATION OF INTEREST GROUPS SURVEYED

Description of Survey

From the literature it was not possible to establish in detail
exactly the role that interest groups have played in influencing
hazard control policy in the areas considered in this study.
In an attempt to fill this apparent gap in the literature and
to provide a uniform basis for comparing the role and
organisation of the interest groups concerned, a survey was made
of a representative sample of groups. Table I lists the groups
selected for survey and their Wootton classification. The survey
was made by obtaining answers to a questionnaire that was
designed to establish the nature of the resources, goals,
internal policy making procedures, interactions with proximate
policy makers, influence the group had had, and the group's
views on probability techniques for determining the acceptability
of particular hazards. Full details of the survey are given in
Appendix V and in the following an analysis is given of the data
the survey produced.

The form of analysis that is adopted is to identify, what
Reynolds calls (170), the interacting variables that are
associated with the evolution of policy and to try and relate
the way these variables are distributed to the role the various
interest groups play in influencing hazard control policy. In
the terminology used by Reynolds these variables are: goals,
influences or capabilities, means or instruments and processes.
To prevent misunderstanding about the way the terms are used in
the analysis they need to be defined. Goals are taken to be
the objectives that the groups aim to achieve. The goals
pursued may be long term or intermediate goals. The long term
goal may be some remote objective that is unlikely to be
achieved in an imperfect world, while the intermediate goals are
those that can be achieved given the incremental nature of
policy development. The capability of a group is taken to be
the sum total of its resources. This includes: technical
knowledge, financial strength, size, established position of the
group, and the nature of people it represents. The means and
instrument variables are the formal and informal contacts a
group enjoys with the proximate policy makers. Finally the
processes are taken to be the parts of the policy making system
that a group has to consider and interact with when it attempts
to influence policy. It also includes the consideration of the
groups own internal structure and policy making procedure.

The interest groups are first considered under their Wootton
classification, and then their significance related to the five
activities is considered. The Interest Groups considered have
the following Wootton classifications: Economic Interest of
the third and second order, Integrative Interests of the third
order and Cultural Interest of the third order.

TABLE 1

INTEREST GROUPS SELECTED FOR SURVEY

AND THE CLASSIFICATION

ACTIVITY	INTEREST GROUPS SELECTED FOR SURVEY	WOOTTON CLASSIFICATION OF INTEREST GROUPS
ROAD TRANSPORT	British Roads Federation Society of Motor Manufacturers and Traders* Motor Industry Research Association* Institution of Mechanical Engineers* Institution of Civil Engineers Road Operators Safety Council Royal Society for the Prevention of Accidents* Automobile Association* Royal Automobile Club* Centre for Study of Responsive Law Motoring Which (Consumers Association)*	) Economic Interest Groups) of Third Order) Economic Interest Group of Second Order) Integrative Interest Groups) of Third Order)) Cultural Interest Groups) of Third order))
AIR TRANSPORT	Society of British Aerospace Companies* Lloyds Aviation Underwriters Association* Royal Aeronautical Society* Guild of Air Traffic Control Officers* British Air Line Pilots Association* Air Registration Board* Flight Safety Committee*	Economic Interest Group of Third order Economic Interest Group of Second order) Integrative Interest Group) of Third order) Economic Interest Group of Second order Integrative Interest Group of Third order Cultural Interest Group of Third order

TABLE 1
(Continued)

ACTIVITY	INTEREST GROUPS SELECTED FOR SURVEY	
FACTORIES	Confederation of British Industry*	Economic Interest Group of Third Order
	Amalgamated Union of Engineering and Foundry Workers*	Economic Interest Group of Second order
	Institution of Mechanical Engineers*	Integrative Interest Group of Third order
	Royal Society for the Prevention of Accidents*	Cultural Interest Group of Third order
NUCLEAR POWER REACTORS	Central Electricity Generating Board*	Economic Interest Group of Third order
	Nuclear Plant Contractors (TNPG and BNDC)) Institution of Professional Civil Servants*) National Union of General and Municipal Workers*)	Economic Interest Group of Second order
AIR CONTAMINATION	Confederation of British Industry*	Economic Interest Group of Third order
	British Medical Association*) Institution of Civil Engineers*) Institution of Public Health Inspectors) Association of Public Health Inspectors*)	Integrative Interest Groups of Third order
	Urban District Councils Association) Association of Municipal Corporations*)	Integrative Interest Groups of Third order
	National Society for Clean Air*) Clean Air Council*) Civic Trust*)	Cultural Interest Groups of Third order

Notes (1) The Conservative*, Labour and Liberal parties were also included in the survey

(2) Groups giving some response to questionnaire marked with *

Economic Interest Groups

The Interest Groups that were classified as Economic Interest
Groups of the third order and which responded to the
questionnaire were: the Confederation of British Industry, the
Society of Motor Manufacturers and Traders, the Society of
British Aerospace Companies, and the Central Electricity
Generating Board. Each of these groups plays many roles besides
the role they play in relation to hazard control policy, so
their role in this respect may be regarded as a by-product of
their other activities. These groups represent a particular
section of society which can more accurately be described as an
elite section rather than a mass section. The main goals of
these groups are related to promoting the industry which the
organisations represent in the case of the trade associations,
and in the case of CEGB to provide a national electricity supply.
The Groups' goal in relation to the development of hazard
control policy seems to be greater reliance on voluntary control
and self regulation. Each of these groups appears to have
fairly extensive financial resources, and has extensive
technical resources either in its own staff or on call from
members' organisations. The groups all actively interact with
proximate policy makers at the formal and at the informal level.
This interaction with proximate policy makers is sought both by
the government and by the groups themselves, so it can be seen
that approval of their role by government is implied by
acceptance of their interaction.

The sample of interest groups classified under the heading of
economic interest group of the second order is rather more
complex. It consists of: The Motor Industry Research
Association, the British Airline Pilots Association, The
Amalgamated Union of Engineering and Foundry Workers, The
Institution of Professional Civil Servants, The National Union
of General and Municipal Workers, and the Lloyds Aviation
Underwriters Association. All these groups, as their prime
function, provide some service to their members other than any
function they perform in an attempt to influence hazard control
policy. The role that they play with regard to influencing
hazard control policy is essentially a by-product of their
other functions. Within the sample there are two slightly
different types of group, they are: the union type, and the
trade specialist service type of group. The latter being
represented by the Motor Industry Research Association, and the
Lloyds Aviation Underwriters Association. These groups are not
strictly mass organisations, although the union type of
organisation does have a very large individual membership the
members are all associated with a particular occupation. On the
other hand they are a rather different type of elite to that
represented by the groups classified as economic interest groups
of the third order. The primary goals of both groups are
broadly similar: to promote the economic welfare of their
members. There is, however, a difference in the concern the
groups have for influencing hazard control policy. The union
type of organisation tends to aim for stricter government
control of hazards, while the specialist service groups seems to

tend to prefer to let market forces dictate the action required to control hazards. The resources on which the capability of these groups is based are fairly extensive in the financial and technical terms, although the Lloyds Aviation Underwriters Association is very much smaller than other members of the sample. There is considerable variation in the extent of the interactions that these groups have developed with proximate policy makers. The range of contacts varies from very well developed formal and informal contacts as in the case of the British Airline Pilots Association to very weak contacts as in the case of Lloyds Aviation Underwriters Association. The group whose members are mainly government employees; the Institution of Professional Civil Servants feels that the extent to which it can attempt to influence policy is in some degree inhibited by the fact that their members are government employees. The union type of organisation tends to deal directly with employers on matters related to the hazard control they are concerned with. The issues on which they deal directly with the proximate policy makers tend to be slightly smaller issues than those taken up by the third order groups. One major exception to this rather sweeping generalisation is the British Airline Pilots Association which is willing to bring the expert technical knowledge of its members to the attention of proximate policy makers and to use strike-type action to emphasise its views. The Association appears to have a very active membership, and many pilots give their spare time to preparing technical papers to support the Association's proposals.

Integrative Interest Groups

Now considering the groups that make up the sample of Integrative Interest Groups of the third order, namely: the Institution of Civil Engineers, the Institution of Mechanical Engineers, the Royal Aeronautical Society, the Air Registration Board, the British Medical Association, the Association of Public Health Inspectors, The Guild of Air Traffic Control Officers and the Association of Municipal Corporations. These groups are to some extent all concerned with: maintaining and promoting some form of standard in their own special area* Like all the other groups so far considered, they perform some service for their members other than any attempt they make to influence policy on hazard control, so their influence on this type of policy can be described as a by-product of their other activities. Two groups, the Royal Aeronautical Society** and

* The Air Registration Board is concerned with aircraft safety standards. The Association of Municipal Corporations is broadly concerned with policy standards in the Corporations. The other organisations are concerned with standards in the professions they represent.

** The Royal Aeronautical Society was requested to, and did, present evidence to the Edwards' Committee, see page

the Institution of Mechanical Engineers* do not, in general,
attempt to influence policy so they could be described as latent
groups. The membership of these groups can fairly be described
as elites, as members have either to go through some form of
selective election process, or else they have to be qualified
in some special way. Apart from: the Association of Municipal
Corporations and the Air Registration Board, the main goal of
these groups can be broadly expressed as being to promote in
their own field the study and acquisition of knowledge for the
public good. The Air Registration Board's goal was to provide
the public with a measure of protection against aviation risks,
and the goal of the Association of Municipal Corporations was to
protect the rights and privileges of corporations. There was no
uniformity of opinion among these groups about their views on
this subject. The general sense of the views expressed appeared
to be that the government should take professional technical
advice on the subject and enforce stricter controls.

The resources that integrative interest groups have are in most
cases quite extensive, and for the larger organisations are
characterised by a membership of over 40,000 and an income of
the order of £500,000 (at 1971 prices). There is a tendency in
these organisations for a large proportion of their members not
to take an active interest in the policy developed and pursued
by the executive body. All these organisations could call on
the extensive technical capabilities of their members to
supplement the capabilities of the permanent staff. The extent
that demands are made on members is roughly proportional to the
extent that the group tends to interact with policy makers being
least in groups like the Institution of Mechanical Engineers*
and Royal Aeronautical Society and greater in organisations like
British Medical Association and the Association of Municipal
Corporations.

The British Medical Association, the Association of Municipal
Corporations, and the Air Registration Board have extensive
formal and informal contacts with proximate policy makers, mainly
because they cover a very wide spectrum of interests for their
members, which necessitates frequent contact with proximate
policy makers. The other integrative interest groups tend to
have mainly informal contacts and to develop opinions on
particular subjects only when their advice and opinions are
sought. It is a characteristic of these groups that those
organisations most concerned to have their views put to the
policy makers have members of parliament and members of the
House of Lords elected to their executive or policy forming body.

---

*   The Institution of Mechanical Engineers has pointed out that since the
    questionnaire was completed a number of contacts with government have
    developed. This shows how the role of an interest group may change
    from time to time.

These groups are really concerned with about the same width of
interest in their own particular field as the economic interest
groups of the third order considered. The main difference
appears to be in the purpose of the interaction they have with
the policy making process. The professional learned society
type of group, such as the Institution of Civil Engineers is
more associated with giving straight technical advice on policy
matters. Groups like the British Medical Association and to a
lesser extent the Association of Municipal Corporations are
sometimes concerned to react with the policy making process in
a way that stimulates the development and adoption of new
policies.

Cultural Interest Groups

The cultural groups of the third order that were studied were:
The Royal Society for the Prevention of Accidents, The Royal
Automobile Club, The Automobile Association, The Flight Safety
Committee, The National Society for Clean Air, The Clean Air
Council, The Civic Trust, and Motoring Which. This section of
the sample of groups was fairly evenly divided between what
could be called special-interest groups and by-product groups.
The groups that could be considered to have an interest in
influencing policy on the control of hazards as a by-product
of their other interests were: The Royal Automobile Club, The
Automobile Association, The Civic Trust and Motoring Which.
Apart from the Flight Safety Committee and the Clean Air
Council all these groups are what could be considered as mass
groups, that is there is very little in the way of a
qualification requirement that has to be satisfied for people
to become members. In contrast the Flight Safety Committee and
the Clean Air Council are really elite groups as election to
membership is to some extent selective.

The long term goals of cultural groups tend, more than in the
other groups considered, to be related to increasing the
enjoyment that the public can derive from the activity the
group is associated with. With a few exceptions this is
manifest in the short term goals of the groups being to bring
about more stringent controls of the hazards associated with the
activities they are interested in. The exceptions to this wish
for greater restriction are represented by the motoring groups
resistance to speed limits on motorways and random breathalyser
tests.

In general the financial and technical resources at the command
of cultural interest groups is of a lower order than that of
the other groups considered although the groups associated with
motoring are not so much lower. These groups seem to have a
more highly developed public relations content in their approach
to influencing policy.

Both formal and informal contacts are well developed with the
cultural groups. It appears that considerable care is often
taken to ensure that parliamentary personalities from both
Houses are elected as vice-presidents, probably in the hope that

they will support the interests of the groups in parliament.
They also attempt to react across the whole spectrum of the
policy making process.

Political Parties

The political parties do not fit under the Wootton
classification, but from the very limited response of the
political parties to the questionnaire it was not possible to
form a balanced and representative view of the role they play in
influencing policy on the control of hazards. Perhaps in the
context of the study the most that can be said is that they act
both as a means for collecting and testing the views of the
electorate and a vehicle through which members of the
electorate can make their views known to the proximate policy
makers.

Relation Of Interest Groups To Activities Studied

Having briefly considered the interest groups in relation to
their Wootton classification, they will now be considered in
relation to the five activities that form the base of this
study.

The first question to be considered is how does the role of the
interest groups relate to the control of hazard policy in the
five activities selected as case studies. To answer this
question the significance of the role that interest groups play
is examined in relation to the size, technical content and age
of the activity.

Interest Groups and Road Transport

Road transport is a very large complex activity, which touches
in some way practically every member of society. None of the
interest groups studied can claim to represent all parties
associated with road transport; the Society for Motor
Manufacturers and Traders represents the motor industry while
the Automobile Association and Royal Automobile Club between
them represent the majority of private road users. The
contribution that the interest groups make towards policy is in
the technical detail of the policy adopted and the timing of its
introduction. The arguments presented by the Society of Motor
Manufacturers and Traders are aimed at, as far as possible,
making industry responsible for the voluntary control of its
affairs and achieving legal performance requirements with the
minimum of disruption of production rather than introducing new
design requirements.* The Motoring Associations on the other

* In this context a performance requirement is some condition such as:
  maximum noise level or exhaust gas specification which the vehicle has
  to satisfy and a design requirement is the specification of the way a
  particular problem is to be solved.

hand are concerned to keep the restrictions on the motorist to
a minimum. The role of the Royal Society for the Prevention of
Accidents, the Institution of Civil Engineers, and Motoring
Which is in a slightly lower key, but it would be misleading to
say that their role is limited to the collection and
dissemination of relevant technical information because the way
they bring views to the attention of people in the policy making
chain does have some influence on the ultimate policy. The
Institution of Mechanical Engineers is really a latent group,
it neither seeks to present its views nor are the Institution's
views sought.*

Road transport was developed for many years with relatively
light controls, consequently after the Second World War when
rapid growth of road usage focussed attention on the need to
improve road safety there were many established practices and
conventions that required modification. The improvements in
vehicle and road design have been brought about in what can best
be described as an incremental manner. Neither the policy
makers nor interest groups have suggested more than an
incremental approach to the problem. The length of time
required to build better roads, design and build safer cars,
and the time taken to change driving habits dces mean that the
timescale for solving today's problems of road transport is
measured in years. Present day problems were not anticipated
in the formative years of road transport, with the result that
changes now seem to be necessary to modify the inherited
standards.

Interest Groups and Air Transport

Air transport is a more modern and compact activity, which has
developed in a climate in which everyone has been aware of the
hazards involved. The interest groups surveyed were the
Society of British Aerospace Companies, Lloyds Aviation
Underwriters Association, the Royal Aeronautical Society, the
Guild of Air Traffic Control Officers, the British Airline
Pilots Association, the Air Registration Board, and the Flight
Safety Committee. These groups probably represent a greater
proportion of the groups concerned with this activity than the
proportion considered in relation to other activities. The
three interest groups that dominated the air transport hazard
control policy influencing scene are the Society of British
Aerospace Companies, British Airline Pilots Association, and
the Air Registration Board. These three groups represent
respectively industry, the pilots and the government.** The
ordinary fare paying passenger is not represented directly

* See note page 90

** The responsibilities of the Air Registration Board were transferred to
   the Civil Aviation Authority on 1 April 1972. The Civil Aviation
   Authority was established by an Act of Parliament, so there is now a
   formal link with the Government.

through these groups, although doubtless each would claim some
concern for the customer. These three groups appear to work
for stricter design standards and tighter operational controls
to reduce hazards through a deeper technical understanding of
the problems. The group that has a unique character with no
direct parallel is the British Airline Pilots Association. This
organisation has by voluntary effort brought the technical
knowledge of flight deck duties to the attention of proximate
policy makers in a way that makes an impact on the policies
developed. The Air Registration Board operated essentially in
the technical mode, seeking the views of a cross-section of the
air transport industry's technical authorities. Now that the
Air Registration Board has become the Airworthiness Requirements
Board of the Civil Aviation Authority it could be argued that it
no longer represents an independent interest group, concern on
just this aspect of the role of the Board was expressed before
the change was made. The other groups considered appear to be
of rather less significance, and appear not to fully exploit
their potential for influencing policy.

The main developments in air transport have taken place in a
period when there has been growing concern to keep hazards to a
minimum, and the main interest groups associated with this
activity appear to have worked consistently towards this end.

Interest Groups and Factories

Factories, in contrast to air transport, represent a large
activity with a very long history, and many individually owned
operational units. Clearly there are many trade associations,
unions and professional organisations associated with the
activity that can be termed factories. The four groups that
responded were the Confederation of British Industry, the
Amalgamated Union of Engineering Workers, the Institution of
Mechanical Engineers and the Royal Society for the Prevention of
Accidents. The Confederation of British Industry is the
organisation that in many fields presents the co-ordinated
views of its supporting trade organisations. It tends not to
get involved in the specialist problems of the road transport
and aircraft industries, but to leave these to relevant
Associations such as the Society of Motor Manufacturers and
Traders, and the Society of British Aerospace Industries. The
Amalgamated Union of Engineering Workers could be considered as
typical of the large unions that represent the workers in
factories. The views that seem to predominate are similar to
those identified for road transport, namely employers would like
to see safety controls developed on a voluntary basis, while
the union view was that stricter statutory control was required.
The role of the Royal Society for the Prevention of Accidents
was again secondary, perhaps because they have no direct
operational role on either the employers' or the employees' side.
The Institution of Mechanical Engineers is to a large extent a
latent organisation in this context. * The groups that do make

* See note page 90

contributions to the formation of policy tend to make their
contributions in the form of specialist advice on the technical
details.

Factories were in existence many years before there was
legislation to try and control the associated hazards.
Although controls were introduced in the last century the
development of these controls has not reached the advanced
level that exists in other activities. Awareness of the
detailed control possible in other activities such as the
nuclear industry and air transport has had some impact on
thinking about factory safety, and the way groups would like
this thinking implemented will be shown later in the review of
evidence presented to the Robens' Committee.

Interest Groups and Nuclear Power Reactors

Nuclear reactors are the newest and the smallest of the five
activities considered and is the activity that is subject to
the tightest government hazard control. It is rather
unfortunate that full replies to the survey questionnaire were
only obtained from the two union type interest groups, the
Institution of Professional Civil Servants and the National
Union of General and Municipal Workers. On the other hand the
control of hazards in this field appears to be regarded as
satisfactory, the Robens' Committee (171) suggested one
advantage of their proposed unified Authority for the control
of Safety at Work was that it would bring to bear on a wider
field the advanced methods of safety and reliability analysis
developed in the nuclear industry. The view expressed by the
National Union of General and Municipal Workers is that they
would like to see even stricter control of hazards associated
with nuclear reactors. The Institution of Professional Civil
Servants feels inhibited in the action they can take as their
members are civil servants and part of the government system.
However, they seem to be fairly satisfied with the way hazards
are controlled in the nuclear industry as they proposed a
similar form of control for general application to the Robens'
Committee. In an attempt to show the role that other interest
groups have played in the nuclear safety field, the positions
the nuclear power plant constructors, the Nuclear Installations
Inspectorate and the United Kingdom Atomic Energy Authority took
up in relation to reactor siting is described in Appendix V.
It shows how within a few years of the United Kingdom Atomic
Energy Authority suggesting that probability techniques could
be used to evaluate the acceptability of particular reactors,
the technique was accepted by the other bodies. Quite a lot of
the movement towards a common view seems to have taken place as
a result of open and public discussion of the problem at
symposia and technical meetings. One final point on interest
groups associated with nuclear reactors, is that in Britain
there does not appear to be any strong well-developed group
arguing that existing nuclear reactors are unsafe, as there is
in other countries such as America, Germany and Switzerland.
This may be due to the fairly open way nuclear power has been
introduced in this country or to the fact that in the past

critics have been able to concentrate their attack on economic
advantages of other power sources.

Interest Groups and Air Contamination

Air contamination is the fifth and oldest of the hazard areas
considered. The Confederation of British Industry, British
Medical Association, Institution of Civil Engineers, The
Association of Public Health Inspectors, the National Society
for Clean Air and the Association of Municipal Corporations
replied to the survey questionnaire and demonstrated that they
had been active in discussions with proximate policy makers
about air contamination and pollution generally.

The Confederation of British Industry's views on pollution
generally were: that a flexible approach should be adopted,
that each case should be judged on its merits, and that British
practice should not be stricter than that in other countries.
The Confederation was also against taxation as a means of
limiting pollution. The other groups generally held the view
that stricter control of air contamination was required. The
British Medical Association and the other integrative groups
direct their attention to the technical aspects of the problem
and in general have only been able to make slow progress towards
a reduction of this form of hazard. The cultural group
considered, the National Society for Clean Air tends to deal
rather more with the problem from the publicity* angle rather
than the technical side and hope that by making more people
aware of the problem improvements will be brought about.

The main problem in reducing air contamination is that in many
cases the sources of air pollution have been established for a
long time, and their use is perpetuated by hallowed conventions.
It was shown earlier how the Alkali Inspectorate finds it
difficult to resist applications to delay the introduction of
stricter controls. This really illustrates that it is easier
and more effective to introduce hazard controls as soon as an
activity is established, and that interest groups find it hard
to get new controls introduced quickly into long established
activities.

Interest Groups and the Robens' Committee

An indication of the way various interest groups respond to a
particular situation in which their views are sought by
proximate policy makers is given by the way they responded to
the Robens' Committee. The Robens' Committee was established,
as already explained, to review and make proposals for change,
if necessary, to the way in which a wide range of industrial
hazards were controlled. The Committee received submissions

* The National Society for Clean Air make the point that their Technical
  Committee takes considerable trouble to ensure that the arguments they
  put forward are technically accurate, and that the solutions they
  suggest are technically feasible.

from 183 organisations and individuals (172). Seven of the
organisations considered in this study submitted evidence to
the Committee, and the evidence of six of them was considered
to be sufficiently apposite to merit inclusion in the selection
of evidence printed in Volume 2 of the report (173).

The role of a proximate policy forming body, such as the Robens'
Committee, is to evaluate all the evidence and advise the
government on the policy it considers should be followed, in
the best interests of the nation as a whole. It follows
therefore that the arguments of the interest groups which the
Robens' Committee incorporated in their report would be those
that they perceived as being in the best interest of the nation
as a whole. Before examining the way the various groups
responded to the Robens' Committee, the findings of the
Committee are examined. In the first chapter of the report of
the Robens' Committee (174) the view was expressed that the
toll of death, injury, suffering and economic waste from
accidents at work and occupational diseases was unacceptably
high. It was suggested that the way to bring about a
progressive improvement was to use deliberate pressures to
stimulate more sustained attention to safety and health at work.
It was noted that there was a lack of balance between the
regulatory and voluntary elements of the overall 'system' of
provision for safety and health at work. The primary
responsibility for doing something about present levels of
occupational accidents and diseases lies with those who create
the risks and those who work with them. The present approach
tends to encourage people to think and behave as if safety and
health at work were primarily a matter of detailed regulation
by external agencies.

The Committee was very critical of the present arrangements and
the following extract from the report (175) seems to summarise
their views:- "Present regulatory provisions follow a style
and pattern developed in an earlier and different social and
technological context. Their piecemeal development has led to
a haphazard mass of law which is intricate in detail,
unprogressive, often difficult to comprehend and difficult to
keep up to date. It pays insufficient regard to human and
organisational factors in accident prevention, does not cover
all work people, and does not deal comprehensively and
effectively with some sources of serious hazard. These defects
are compounded and perpetuated by excessively fragmented
administrative arrangements.

A more effective self-regulating system is needed. Reform
should be aimed at two fundamental and closely related
objectives. First, the statutory arrangements should be
revised and re-organised to increase the efficiency of the
state's contribution to health and safety at work, and
secondly, the new statutory arrangements should be designed
to provide a framework for better self-regulation."

The main proposal of the Committee was that a National Authority
for Safety and Health at Work should be set up, and that

present safety and health legislation dealing with factories, mines, agriculture, explosives, petroleum, nuclear installations and alkali works should be revised, unified and administered by the new Authority. The Authority should have a distinct, separate identity, with its own budget, and full operational autonomy under the broad policy directives of a departmental Minister. It should have a comprehensive range of executive powers and functions. Statutory provisions formulated by the Authority should be laid before Parliament by the sponsoring Minister.

The other important features of the report are as follows:

1.   It was suggested that employers should be required to set out written statements of their safety and health policy and provisions, and that companies should employ systematic prevention techniques (that includes the systematic analysis of all operations, plants and processes). Work people should be involved in the arrangements for monitoring safety and health arrangements, and there should be a general statutory obligation on employers to consult with their work people on measures for promoting safety and health.

2.   Attention was drawn to the need for a better mechanism for linking up the efforts of industry-level safety bodies with the work of the statutory services.

3.   The view was also expressed that voluntary standards and codes of practice provide the most flexible and practical means of promoting progressively better conditions of safety and health at work, and that wherever possible they should be used in place of statutory regulations.

4.   As a matter of explicit policy, the provision of expert and impartial advice and assistance to industry should be the basic function of the unified inspectorate. At the same time, tighter control over serious problems should be exercised through the more effective deployment and use of inspection personnel.

5.   The work of local authorities related to safety and health should be co-ordinated and integrated with the work of the area offices of the new national authority.

6.   It was recommended that new legislation should be so formulated as to ensure that the interests of the public as well as employees are taken fully into account.

7.   It was recommended that there should be a statutory obligation on the makers and vendors of plant and equipment to ensure that the plant and equipment complied with all safety provisions, and that the new National Safety Authority would have power to make special regulations concerning safety design and construction.

8.   It was recommended that the Employment Medical Advisory Service, to be brought into operation by the end of 1972, should function as part of the Authority for Safety and Health at Work, and should maintain close operational liaison with the National Health Service.

9.   It was proposed that there should be improved training facilities in safety, and a more co-ordinated research effort in occupational safety and health.

Six groups considered in this study appear to have influenced the views formed by the Robens' Committee. The groups were: the Association of Municipal Corporations, the Association of Public Health Inspectors, the British Medical Association, the Confederation of British Industry, the Institution of Professional Civil Servants, and the Royal Society for the Prevention of Accidents, and their evidence was published in Volume 2 of the Committee's Report.

The contribution that the groups were expected to make was suggested by the memorandum issued by the Committee to assist organisations wishing to submit evidence. The memorandum set out the range of interest of the Committee, and asked a number of questions that were grouped under the headings: legislation, administration and enforcement, voluntary effort, safety and health at the work place, training, design, overseas experience, compensation, statistics, research, public safety and costs. The headings used for grouping the questions, are also the heading under which the Committee finally made its report.

The main features of the submissions that the six organisations made in their published evidence are as follows:

The Association of Municipal Corporations suggested (176) that local authorities should have greater responsibility for administering health and safety legislation, that voluntary effort to improve safety should be encouraged, and that hazards to the public needed further consideration. The suggestion that was not accepted without reservation by the committee was that local government should have more responsibility for administering health and safety legislation. While the Committee appreciated that local government should have an increasing role it was concerned about the unevenness in the quality of local authority inspection. Views were expressed, as already mentioned, that the contribution of the local authorities could be used to full advantage by having it co-ordinated and integrated with the work of the area offices of the new National Authority.

The Association of Public Health Inspectors advocated (177) unification of safety legislation, proper training in occupational safety and health, a greater role for local authorities in the administration of health and safety legislation, supported voluntary effort to improve safety, all machinery and equipment sold should satisfy safety standards, and that protection of the general public from hazards and

nuisance arising out of industrial activities is a subject
which requires investigation. All the suggestions made by
the Association were incorporated in the Committee's report
with the exception of the greater role for local authorities
in the administration of health and safety legislation. The
view the committee took on this last point is described above,
in relation to the similar proposal made by the Association of
Municipal Corporations.

The submission put forward by the British Medical Association
(178) was mainly concerned with describing the occupational
health service, that it considered should be established,
although it also contained a suggestion for improved training
in occupational medicine, and recommended that legislation was
required to protect the public against increasing pollution from
noise and exhaust fumes. In commenting on the proposals for an
occupational health service the Committee expressed the view
that it was a proposal that extended well beyond the Committee's
terms of reference, and that it was a subject that raised
fundamental issues concerning the deployment of national
medical resources. It was also postulated that the issues
involved were very broad and long term and that a proper
evaluation would require an extensive analysis of the costs and
benefits of the possible alternative forms of organisation.
However, it is interesting to note that the suggestion of an
Occupational Health Service was included in the Labour Party's
programme (179), perhaps this was a by-product of the BMA's
presentation to the Robens' Committee. The sense of the
proposals related to training and protection of the public was
incorporated in the recommendations made by the Committee.

The evidence that the Confederation of British Industry
presented (180), was one of the longer submissions, running to
over forty pages. The evidence set out details of how
legislation could be unified, proposed unification of the
various inspectorates, recommended better training of inspectors
and more safety training generally, proposed greater reliance
on voluntary effort to improve safety, endorsed the concept
that self-certification of products as the preferred method of
approving goods for use, and argued that unification of the
inspectorates would make it easier to co-ordinate efforts to
improve public safety. Only the emphasis on self-certification
of products was not reflected in the proposals put forward by
the Committee.

The Institution of Professional Civil Servants proposed (181)
that safety legislation should be co-ordinated and unified,
there should be a co-ordinated enforcement and advisory service
in one Ministry, inspectorates should be properly staffed,
legal sanctions should be more widely used, training improved
and co-ordinated research facilities should be available. As
voluntary arrangements cannot replace legislation, legislation
should be passed to empower inspectors to apply for court orders
where the safety of the public is endangered by radiation or
toxic waste. In general the views of the Institution were close
to the proposals made by the Committee, although there are

slight differences in emphasis, particularly with regard to the
extent that legal sanctions should be used, and the reliance
that should be placed upon voluntary methods to improve safety.

The Royal Society for the Prevention of Accidents suggested (182)
simplification of the law, an increase in the number of
inspectors, endorsement of voluntary methods of improving safety
improved training, safety should be built in at the design
stage, and more research on safety was required. In general the
proposals put forward by the Society were less radical than the
reforms put forward by the Committee.

By comparing the evidence presented with the proposals of the
Committee it is, even with the small sample of organisations
considered, possible to see that where technically feasible
proposals are put forward by several groups their opinion is
likely to be accepted. This suggests that the Committee
attempted to form a view based on the consensus of opinion found
in the evidence presented to it. The proposal made by the BMA
for an occupational health service was not accepted because it
was considered to be a proposal outside the terms of reference
of the Committee, also it was a proposition that was not fully
argued in terms of cost and benefits to the community as a
whole. From the BMA's point of view it may have given the
proposition a useful airing as it seems to have been taken up
by the Labour Party.

The indications are that the more technically justified
arguments presented to the proximate policy makers will be
accepted if he perceives them to be representative of the
needs of the community as a whole. It does not follow that it
is only the largest organisations that can put forward a
comprehensive case that finds acceptance. Although it has been
indicated that much of the very detailed evidence put forward
by the Confederation of British Industry was accepted by the
Committee. Private individuals can still have a considerable
impact on the policy proposed by proximate policy makers. An
example of this is to be found in the evidence submitted to
the Robens' Committee.

The paper presented to the Committee by Mr. A. Cook (183)
contained detailed proposals for a national centre for
safeguards and reliability, which were very similar in concept
to the proposal that the Committee developed for what it called
a National Authority for Safety and Health at Work. Mr. Cook
is a member of the Safety and Reliability Directorate of the
United Kingdom Atomic Energy Authority and has many years of
executive experience in nuclear safety work. It appears
reasonable to assume that it was because Mr. Cook's proposals
were based on practical experience in safety work, were not
biased to personal advantage, and appeared to represent the
needs of the community, that they were considered sufficiently
important by the Committee to warrant publication. This
suggests that the views of an individual are acceptable when
they are seen to represent the views of a section of the
community.

The evidence presented to the Robens' Committee shows how it is the detailed technical knowledge of the activity the interest groups are associated with that is sought by the proximate policy makers. The terms of reference under which the proximate policy maker works tends to act as a constraint on the extent of the influence an interest group can have on policy formation.

THE ROLE OF INTEREST GROUPS IN RELATION TO THE FORMATION OF HAZARD CONTROL POLICY

From the above examination of interest groups it is possible to suggest that a proximate policy maker's perception of the usefulness of an interest group might be based on the extent that an interest group's representations satisfy the following criteria:-

1.   Proposals put forward are based on a thorough technical understanding of the problem.

2.   Proposals are such that they can be implemented within the present framework of government.

3.   Proposals are in line with the need of the community as a whole, and are not just representative of a sectional interest.

The survey also suggests a number of general conclusions about the role interest groups play in influencing policy on the control of hazards. The conclusions are as follows:-

1.   In general it does not appear that the interest groups are solely responsible for the differences in the method of hazard control that have been adopted in the five activities considered. It seems more likely that the differences are due to the different ages of the activities, differences in the understanding of the relevant technology, opinions on the form of control that could be adopted at the time the controls were first introduced. The most interest groups appear to be able to achieve is delay in the introduction of a control or detail technical modification in the form of control. The delay some trade associations have brought about in the restriction of gaseous discharge has been shown to have caused concern.

2.   When an interest group is actively trying to influence policy the interactions that take place are broadly as indicated in the diagram of the hazard control policy making sub-system shown in Fig.3

3.   The role that an interest group can assume is limited by its resources and the constraints that the policy makers place on it by way of recognition and specifying the timing and content of interactions permitted.

4.   It is mainly detailed and technical knowledge of the activity the interest group is associated with that the proximate policy maker seeks from the interest group.

5.   It is the membership of a group that decides whether a
group is an active or latent interest group. This was
demonstrated particularly by the fact that of the several
similar integrative groups considered some actively tried to
influence policy while others were quite latent.

6.   The most active groups were generally the economic groups
which took an interest in influencing policy cn hazard control
as a by-product of their other activities. It may be that
because they had well developed internal organisations to
perform their prime function they were better able to take
advantage of the opportunities offered to influence hazard
control policy.

7.   The Robens' Committee report showed how widely proximate
policy makers consult interest groups when they are building up
and confirming their views on the advice they should give on
future policy. This convention, in the hazard control field,
of consulting a wide range of interest groups, suggests that
the proximate policy makers attempt to find the common
denominator in interest group views, so that the view of no one
interest group predominates. This also suggests that wide
consultation, such as takes place in the hazard control field,
limits the power of one interest group.

8.   The trade association type of group in general seems to
aim for control on a voluntary basis, while the union type of
group and the integrative and cultural types of group seem to
prefer compulsory controls more. This suggests that to a
considerable extent a trade associations interest in hazard
control policy formation is to protect the economic interests
of their members.

9.   If the resources of an interest group are limited in
some way such as in their depth of technical knowledge or their
interaction with policy makers is limited because their members
are civil servants then the role the group can play is
proportionally reduced.

10.   The cultural type of interest group are those that mainly
represent the interests of the general public, as opposed to the
specialised public, but apart from the motoring field, these
groups seem to play a less significant role than the other
types of group.

11.   The survey did show considerable differences between the
groups in the extent of their understanding of methods of
evaluating hazards and the controls that can be made effective.

The sample survey of interest groups did show that the Wootton
method of classification gave a useful coarse classification of
groups. However, the study did reveal six additional
characteristics of interest groups, which had some bearing on
the way they attempted to influence hazard control policy, and
these should be identified by any comprehensive method of
classification. The features are: whether the group is active

or latent, whether the group is representative of employers or
employees, and whether the group is a single or multi-interest
group. Recognition of these six features helps considerably
to indicate the type of response that might be predicted from
a particular group. This recognition could be achieved by
adding three terms to the Wootton classification, each of the
three terms defining one of the features just mentioned. With
this refinment of classification the Confederation of British
Industry would be described as "An active employer multi
interest economic group of the third order", while the Royal
Aeronautical Society would be described as "A latent employee,
single-interest integrative group of the third order" and the
Automobile Association would be described as "An active employee
single-interest cultural group of the third order".

It will be seen from these examples that the way the term
"employee" is used is an extension of the normal usage of the
word, but used in this way it is less derogatory than
alternative terms such as mass, which could imply unthinking
members of the public. The advantage of the term "employee"
is that it does clearly differentiate from the term "employer"
which in this study was seen to be groups with slightly
different views on hazard control policy.

CHAPTER 5
CONCLUSIONS

From this study three groups of conclusions can be drawn, they are related to the nature of hazards, hazard control policy and interest groups.

THE NATURE OF HAZARDS

1.   It is doubtful if all technological hazards will ever be completely eliminated, even if technological hazards were eliminated man would still be faced with a great range of natural hazards over which he has no control.

2.   Hazards may arise due to lack of technical knowledge on how to prevent them.

3.   The circumstances that cause a hazard are not always recognised.

4.   Hazards may result simply from people not considering the implications of their actions on others.

5.   Individuals and governments may, for a variety of reasons, consider certain hazard levels acceptable although these levels may not be stated in quantitative terms. The reasons for considering certain levels of hazard acceptable may include a rather tenuous economic argument or an argument on national priorities.

HAZARD CONTROL POLICY

1.   There is no single government statement about hazard control policy covering the five cases considered, but the analysis showed that the policy applied has the following general characteristics:-

>   i   The activity of the community is monitored in a way that it is hoped will identify hazards or changes to hazards as soon as possible.

>   ii   When significant hazards are identified a legal framework is established for their control.

>   iii The control of hazard function is exercised through some form of specialist inspectorate that is empowered and equipped to establish and enforce the necessary controls.

>   iv   To enable interested and informed parties to make their views known to proximate policy makers

consultative machinery is established.

v   The research necessary to solve hazard control problems is undertaken.

2.   The current differences in hazard control policy are more related to the age and size of the activity than to whether the activity is privately or government owned. The older activities are not as tightly controlled as the modern products of technology like aircraft and nuclear power reactors. It is more difficult to apply controls strictly to a large population of moving units like motor cars than to a small number of static units like nuclear power reactors.

3.   The objective of policy to control hazards appears to be to identify the significance of hazards as soon as possible, to keep the burden of hazards within limits considered to be acceptable to the public and to keep expenditure on the control of hazards in harmony with other demands for resources.

4.   The essential characteristics of the hazard control policy making process appear to be:

i   Policy making dominated by the civil service

ii   When policy is being developed use is often made of independent specialist committees and it is through these committees that interest groups can influence policy.

iii   Demands for new policies often arise in the department with the relevant responsibility.

iv   Direct concern of political parties small as compared with defence, economic and foreign policy.

v   Parliamentary scrutiny limited

vi   Cabinet involvement small.

5.   The model of the general policy making system that was postulated suggested policy was formed as the end product of the interaction between public demands, political parties, interest groups, parliament, the Cabinet and the civil service, in an environment conditioned by economic systems, foreign political systems and the current state of knowledge. The model gave a useful guide to the actors and interactions that should be considered in the analysis of the general policy making process. The study showed that hazard control policy making process could be described as a specialist based sub-system of the general policy making system. It is a weakness of the model, and perhaps a weakness of all models, that it gives no indication of the policy likely to develop in a particular situation.

6.   It is possible that future developments in hazard
control policy will include: unification outside the central
government machine of organisations responsible for hazard
control, detailed evaluation in quantitative terms of plants
and products to establish before they are put into use whether
their hazard characteristics are acceptable, greater
protection against acts of sabotage, and all workers to have a
detailed record of the hazards they have been exposed to in the
course of their employment.

INTEREST GROUPS

1.   Interest groups do have a role in the hazard control
policy making system. In broad terms their role could be
defined as being limited to influencing detailed aspects of
policy, they are not the body that finally makes the policy.
The influence that they have is mainly related to: the timing
of the introduction of controls, detailed modification of policy
to make it more acceptable to a particular activity, and
suggesting new ways in which policy may develop. An example of
the latter, is the way the British Medical Association floated
the idea of an occupational health scheme at the Robens'
Committee, an idea that was later taken up by the Labour Party.
Suggestions interest groups put forward for major changes in
policy, such as unification of the various inspectorates
concerned with hazard control, are only taken up if it is seen
that such a view is supported by several groups. The proposals
of the Robens' Committee seem to represent the consensus of
views that interest groups put to them. The role that interest
groups play is to a very large extent dictated by the extent
government seeks the views of interest groups. The convention
in the hazard control policy making process of extensive
consultation with interest groups does appear to significantly
limit the power of any one group.

2.   Interest groups of the third order of economic and
integrative types that concern themselves with hazard control as
a by-product of their other activities seem to have the most
influence on policy. This may be because they have well
developed organisations and contacts developed to satisfy the
needs of their other function.

3.   A more comprehensive classification of interest groups
can be obtained by extending the Wootton method of
classification to indicate the nature of the following
additional features of the group:-

        i   Whether the group is a multi or single interest
        group.

        ii   Whether the group represents employers or
        employees.

        iii Whether the group is active or latent.

1. FINER, S E   Comparative Government. Published
                        Allen Lane. The Penguin Press London
                        1970. pp 62-74

2. WOOTTON, G   Interest Groups. Published Prentice
                        Hall Inc. Englewood Cliffs, New Jersey
                        1970. pp 38-44

3. FARMER, F R   Siting Criteria - A New Approach.
                        A paper presented to the IAEA
                        Symposium on the Containment and Siting
                        of Nuclear Power Reactors held in
                        Vienna from 3-7 April 1967. Published
                        by the United Kingdom Atomic Energy
                        Authority in Atom, No.128, June 1967
                        p.152.

4. OTWAY, H.J and   Reactor siting and design from a risk
    ERDMAN, R C   viewpoint. Nuclear Engineering and
                         Design Vol.13 (1970) No.2 Published
                         by the North Holland Publishing Company
                         Amsterdam. pp 365-368

5.   Annual abstract of Statistics No.107,
                         1970. Published by Her Majesty's
                         Stationery Office 1970. Table 35, p.38

6. BOWEN, J H   Risk from a Super Novae compared with
                         the risk standard for nuclear reactors.
                         Nature Vol.220 October 19, 1968.
                         pp 303-304.

7. PLOWDEN, W   The Motor Car and Politics 1896-1970.
                         Published The Bodley Head Ltd.London
                         1971 pp 456-457

8   Annual Abstract of Statistics. op cit.
                         Table 6 p7.

9. MacMILLAN, R H   Primary Safety: Vehicle Design to avoid
                         Accidents. Proceedings of the
                         Institution of Mechanical Engineers
                         1972. Vol. 186 35/72 pp 479-480.

10.   Road Research 1969. The Annual Report
                         of the Road Research Laboratory.
                         Published by Her Majesty's Stationery
                         Office 1970, pp 44-49

11   Annual Abstract of Statistics op cit.
                            Table 255, p 233.

12.   TYE, W   Airworthiness of the Air Registration
                            Board. The Aeronautical Journal
                            No.719 Vol.74. November 1970, p.885.

13.   BLACK, H C   The Airworthiness of Supersonic
                            Aircraft. The Aeronautical Journal
                            No.686. Vol.72 February 1968 p.118.

14.   NEWTON, E   The Investigation of Aircraft Accidents
                            The Journal of the Royal Aeronautical
                            Society. No.639. Vol.68 March 1964.
                            pp 159-164

15.   LUNDBERG, B K O   Pros and Cons of Supersonic Aviation
                            The Journal of The Royal Aeronautical
                            Society No.645. Vol.68, September 1964
                            p.618

16.   Cmnd 4146   Annual Report of HIM Chief Inspector
                            of Factories, 1968. Published by Her
                            Majesty's Stationery Office 1969 p.72

17.   Cmnd 4146   Op cit. p.74

18.   Cmnd 4146   Op cit. pp 80-84 and 107-109

19.   Cmnd 1225   The hazards to man of Nuclear and
                            Allied Radiations. A second report to
                            the Medical Research Council. Published
                            by Her Majesty's Stationery Office 1960
                            p.5

20   EISENBUD, M   Environmental Radioactivity. Published
                            by McGraw-Hill. New York. 1960
                            pp 135-136, p.392

21.   Cmnd 1225   Op cit. p 22

22   Cmnd 9780   The Hazard to Man of Nuclear and Allied
                            Radiations. Published by Her Majesty's
                            Stationery Office 1956. p.71

23.   Cmnd 1225   Op cit. pp 39-40

24.   Cmnd 1225   Op cit. p 16.

25.   Electrical Review. 17 March 1967.
                            Vol. 180 No.11, p.394.

26.   Central Electrcity Generating Board
                            Annual Report and Accounts 1968-9
                            Published by Her Majesty's Stationery
                            Office pp 84-85

27   BELL, G D   Safety Criteria. Nuclear Engineering and Design, Vol.13, No.2 August 1970. Published by the North Holland Publishing Company, Amsterdam.pp 187-190.

28.   Cmnd 9322   Report of the Committee on Air Pollution. (The Beaver Report) Published by Her Majesty's Stationery Office 1954. pp 8-10.

29.   BUGLER, J   Polluting Britain. Published Penguin Books Ltd. Harmondsworth 1972, p.24.

30.   HECLO, H H   Policy Analysis. The British Journal of Political Science Vol.2 Part 1 January 1972. Published Cambridge University Press, p.85

31.   CHECKLAND, P B   A Systems Map of the Universe. Journal of Systems Engineering, Winter 1971. Vol 2, No.2. Published by the Department of Systems Engineering University of Lancaster, p 110-111

32.   BLACK, G   The application of Systems Analysis to Government Operations. Published F A Praeger, New York. Second Printing 1969. p.107.

33.   ROBERTS, G K   Political Parties and Pressure Groups in Britain. Published Weidenfeld and Nicolson. London 1970. p.134.

34.   WALKER, P G   The Cabinet. Published Jonathan Cape, London 1970. pp 112-121

35.   FOX, J   The Brains Behind The Throne. The Sunday Times Magazine. London 25 March 1973. pp 50-54

36.   PLOWDEN, W   The Motor Car and Politics op cit.p.456

37.   PLOWDEN, W   The Motor Car and Politics op cit.p.22

38.   PLOWDEN, W   The Motor Car and Politics op cit.p.31

39.   PLOWDEN, W   The Motor Car and Politics op cit.p.57

40.   PLOWDEN, W   The Motor Car and Politics, op cit.p.281

41.   PLOWDEN, W   The Motor Car and Politics op cit.p.303

42.   PLOWDEN, W   The Motor Car and Politics op cit.p.281

43   PLOWDEN, W   The Motor Car and Politics op cit.p.56
                           and p.251

44.   WENLOCK, E K   Kitchin's Road Transport Law. 15th
                           Edition. Published Iliffe Books,London
                           1970, p.33

45.   FOSTER, C D   The Transport Problem. Published
                           Blackie & Son Ltd. London 1963. p.167

46.   FOSTER, C D   The Transport Problem op cit.p.168

47.   BUCHANAN, C D   Mixed Blessings. The Motor in Britain,
                           Published Leonard Hill (Books) Ltd.,
                           London 1958. p.113, p.119 and p.125

48.   BUCHANAN, C D   Mixed Blessings op cit.p.132.

49.   Protecting the interest of the motorist
                           A report of the RAC's public policy
                           activities during 1970. Published by
                           the Royal Automobile Club, London.

50.   Motorways Crash Decision; The Economist
                           22 August 1970. p.21.

51.   Pay Now Live Later. The Economist,
                           5 September 1970, p.22

52.   GREGORY, R   The Minister's Line: or the M4 comes
                           to Berkshire. Printed in Public
                           Administration, Summer 1967 p 113-128
                           and Autumn 1967 p.269-286, Vol.45
                           The Royal Institute of Public
                           Administration, London.

53.   WENLOCK, E K   Kitchin's Road Transport Law op cit.
                           p.6, 11-28 and 49-54.

54.   NADER, R   Unsafe at any speed. Published by
                           Grossman Publishers, New York 1965
                           p.30-80

55.   The Economist, 22 August 1970.
                           Published London. p.40.

56.   Now Nader urges recall of "unsafe" VWs
                           Sunday Times 12 September 1971.

57   "Unsafe" British car charge rejected.
                           Daily Telegraph 1 June 1971.

58.   Minister defends British initiative on
                           car safety. Daily Telegraph 11 June
                           1971.

59.   Commons Questions. Plea for study of
                                 car safety plan. Daily Telegraph
                                 17 June 1971.

60.   The Swedish Lesson. The Sunday Times
                                 Magazine 30 May 1971 p.16-23

61   MOT test called a mockery after
                                 garages fail to find 177 faults in 60
                                 studies. The Times 7 October 1971.

62.   Garages reject car test scheme.
                                 Daily Telegraph 16 October 1971.

63.   Not exactly Peyton's Place. Daily
                                 Telegraph 27 October 1971.

64.   The British Imperial Calendar and
                                 Civil Service List 1972. Published by
                                 Her Majesty's Stationery Office London
                                 Col. 287-381

65   PLOWDEN, W   The Motor Car and Politics op cit.
                                 p.370-371.

66.   WRAITH, R E and   Public Enquiries as an Instrument of
          LAMB, E B   Government. Published George Allen &
                                 Unwin Ltd. London 1971.p.343 and p.361.

67.   PLOWDEN, W   The Motor Car and Politics op cit.
                                 p.335 and p.340.

68.   Road Research 1969. Annual Report of
                                 the Road Research Laboratory. Published
                                 by Her Majesty's Stationery Office,
                                 London 1970. p.X

69.   PLOWDEN, W   The Motor Car and Politics op cit.
                                 p.419-420

70.   PLOWDEN, W   The Motor Car and Politics op cit.p.400

71.   PLOWDEN, W   The Motor Car and Politics op cit.p.283

72.   Time to Crack Down on Wild Drivers.
                                 John Peyton, Minister for Transport
                                 Industries talks to Courtenay Edwards.
                                 The Sunday Telegraph. 12 December 1971
                                 p.7

73.   CAPLAN, H   The Law versus Science in Aeronautics.
                                 Journal of the Royal Aeronautical
                                 Society, London. Vol.65, No.607
                                 July 1961, p.453.

74. TYE, W   Airworthiness and the Air Registration
                       Board, op ci.p.877.

75. RADLETT, H G   Airworthiness, the Board and Civil
                       Aviation. The Journal of the Royal
                       Aeronautical Society, London. Vol.69
                       No.651. March 1965 p.176

76. TYE, W   Airworthiness and the Air Registration
                       Board, op cit.p.880

77. LEGG, K   Memorandum to the Committee of Inquiry
                       into Civil Air Transport. The
                       Aeronautical Journal Vol.72 p.695,
                       November 1986 p.965-969. Published by
                       the Royal Aeronautical Society London

78.   Civil Aviation Act 1971. The Public
                       General Acts and Church Measures 1971,
                       published by Her Majesty's Stationery
                       Office London 1972. p.1552-1553.

79.   RAeS talks to Professor David
                       Keith-Lucas. Aerospace December 1972.
                       The monthly newspaper of the Royal
                       Aeronautical Society London, p.4

80. WHEATCROFT, S   Licensing British Air Transport.
                       Journal of the Royal Aeronautical
                       Society London. Vol.58, No.639
                       March 1964, p.171.

81. WHEATCROFT, S   Licensing British Air Transport
                       op cit. p.172.

82. OWEN, D   The "Air Miss" Crisis. The Daily
                       Telegraph Magazine 8 January 1971
                       p.12-15.

83.   Air Traffic Men alarmed by missing
                       links in Heathrow super system. The
                       Sunday Times 14 February 1971, p.2

84.   £25m air control system collisions
                       worry pilots. Daily Telegraph
                       15 February 1971, p.2

85. DONALDSON, E M   Air Traffic scheme unsafe say pilots.
                       Daily Telegraph 26 February 1971.

86.   Her Majesty's Ministers and Heads of
                       Public Departments No.127. December 1972
                       Published by Her Majesty's Stationery
                       Office London p.82-87.

87.   The British Imperial Calendar and
                             Civil Service List 1972. Op cit. col.
                             646-705.

88.   TYE, W   Airworthiness and the Air Registration
                             Board. op ci.p.877-880

89.   TYE, W   Airworthiness and the Air Registration
                             Board. Op cit. p.876 and p.884

90.   Check on pilots hearts. Daily
                             Telegraph 27 February 1973. p.6

91.   A heartbeat away. The Economist.
                             25 November 1972, p.89

92.   SAMUELS, M   Factory Law 8th Edition. Published by
                             Charles Knight London 1969 p.1

93.   Cmnd 4146   Annual Report of HM Chief Inspector of
                             Factories 1968. Published by Her
                             Majesty's Stationery Office,London
                             September 1969, p.XI.

94.   Cmnd 4461   Annual Report of HM Chief Inspector
                             Factories 1969. Published by Her
                             Majesty's Stationery Office London
                             September 1970   p XI-XVII   p 34-48

95.   Hansard Vol.797 No.70 Monday 2 March
                             1970. Published by Her Majesty's
                             Stationery Office London Col.59-62

96.   ROBENS LORD   Human Engineering. Published by
                             Jonathan Cape, London 1970.   p 121-129

97.   WHO'S WHO 1971   Published A & C Black London 1971

98.   COMMONWEALTH   Published by the Association of
       UNIVERSITIES   Commonwealth Universities London 1971
       YEAR BOOK 1971   p.593.

99.   Cmnd 5034   Safety and Health at Work Report of
                             the Committee Published by Her
                             Majesty's Stationery Office London
                             1972 p. XIV-XV

100.   Cmnd 5034   Safety and Health at Work op cit.p.152

101.   Cmnd 5034   Safety and Health at Work op cit.p.153

102.   House of Lords Official Report Vol.338
                             No.31 Tuesday 30 January 1973.
                             Published by Her Majesty's Stationery
                             Office London Col.564.

103.   Hansard Vol.847 No.26 Tuesday
                        5 December 1972. Published by Her
                        Majesty's Stationery Office London
                        Col.1080.

104.   Hansard Vol.847 No.26 op cit. Col.1082

105. Cmnd 5034   Safety and Health at Work op cit.p.156

106. GREENBERG, L   Does Government Enforcement help
                        Industrial Society. Engineering
                        September 1972 p.854-857

107. CRAIG SINCLAIR, T A cost-effectiveness approach to
                        industrial safety. Published by Her
                        Majesty's Stationery Office London 1972
                        p.35

108.   Her Majesty's Ministers and Heads of
                        Public Depts. No.127. December 1972
                        op cit.p.37

109.   The British Imperial Calendar and
                        Civil Service List 1972 op cit.Col.
                        265-285.

110.   Safety and Health at Work. Report of
                        the Committee. Vol.2 Selected written
                        evidence. Published by Her Majesty's
                        Stationery Office London 1972 p.208.

111. BERTIN, L   Atom Harvest. Published by the
                        Scientific Book Club London p.19-22

112. BERTIN, L   Atom Harvest. op cit. p.22-31

113. Cmnd 9389   A programme of nuclear power.
                        Published by Her Majesty's Stationery
                        Office. February 1955. Reprinted 1956
                        p.6

114. STREET and FRAME Law Relating to Nuclear Energy.
                        Published by Butterworths London 1966
                        p.14.

115. STREET and FRAME Law Relating to Nuclear Energy.op cit.
                        p.22

116.   Radioactive Substances Act 1948. The
                        Statutes Third Revised Edition Vol.XXI
                        Published by Her Majesty's Stationery
                        Office London. p.416.

117.   The Public General Acts and Church
                             Assembly Measures of 1954. Published
                             by Her Majesty's Stationery Office
                             London. p 93-94.

118   The United Kingdom Atomic Energy
                             Authority First Annual Report 1954-55.
                             Published by Her Majesty's Stationery
                             Office London p.23

119. EISENBUD, M   Environmental Radioactivity op cit.p.345

120.   Hansard Vol.779 No.72 Thursday 6 March
                             1969. Published by Her Majesty's
                             Stationery Office London Col.626.

121. Cmnd 342   Report of the Committee appointed by
                             the Prime Minister to examine the
                             Organisation for Control of Health and
                             Safety in the United Kingdom Atomic
                             Energy Authority. Published by Her
                             Majesty's Stationery Office London
                             January 1958. p.11.

122.   Hansard Vol.779 No.72, Thursday
                             6 March 1969. Published by Her
                             Majesty's Stationery Office London.
                             Col.628

123.   Extract from the Committee Stage of
                             the Atomic Energy Authority Bill in the
                             House of Lords on 4 February 1971.
                             Atom No.173 March 1971. Published by
                             the United Kingdom Atomic Energy
                             Authority p.61.

124.   Nuclear Safety Advisory Committee.
                             Atom No.138 April 1968. Published by
                             The United Kingdom Atomic Energy
                             Authority, p.88-89

125.   Opening Address by Sir Owen Saunders
                             printed as an introduction to the
                             Report of the Symposium on Safety and
                             Siting. Published by the Institution
                             of Civil Engineers for the British
                             Nuclear Energy Society 1969.

126. BEATTIE, J R and   A Possible Standard of Risk for Large
      BELL, G D   Accidental Releases. Paper IAEA/SM-
                             169/33 at the IAEA Symposium at
                             Julich in February 1973. p.7-9 and
                             Table V

127.   Clean Air Year Book 1970-71.
                           Published by the National Society for
                           Clean Air, Brighton. pp 42-63

128.   The 103rd Annual Report on Alkali and
                           Works by the Chief Inspectors 1966.
                           Published by Her Majesty's Stationery
                           Office London 1967. pp 16-27.

129.   The Air We Breathe. The Daily
                           Telegraph Magazine No.344 28 May 1971.
                           pp 16-27.

130.   The 104th Annual Report on Alkali and
                           Works by the Chief Inspector 1967.
                           Published by Her Majesty's Stationery
                           Office London 1968. pp 1-14.

131   The 105th Annual Report on Alkali and
                           Works by the Chief Inspector 1968.
                           Published by Her Majesty's Stationery
                           Office London 1969. pp 4-11.

132. Cmnd 9322   Report of the Committee on Air
                           Pollution op cit.p.11

133   The 106th Annual Report on Alkali and
                           Works by the Chief Inspector 1969.
                           Published by Her Majesty's Stationery
                           Office London 1970. pp 8-9. pp 17-18.

134. Cmnd 4585   Royal Commission on Environmental
                           Pollution. First Report published by
                           Her Majesty's Stationery Office London
                           1971. p.46-47.

135.   Clean Air Year Book 1970-71 op cit.p.9

136.   Clean Air Year Book 1970-71. op cit.
                           p.12-28.

137. BUGLER, J   Polluting Britain op cit.p.3-31

138. Cmnd 5034   Safety and Health at Work op cit.p.34

139. Cmnd 5034   Safety and Health at Work op cit.p.184

140.   Pollution Nuisance or Nemesis. A
                           report on the Control of Pollution
                           presented to the Secretary of State
                           for the Environment. Published by
                           Her Majesty's Stationery Office 1972.
                           pp 38-40, 58, and 80-83.

141.   Hansard Friday 22 December 1972.
                             Published by Her Majesty's Stationery
                             Office, London. Col. 1793-1795.

142.   Hansard Wednesday 28 February 1973.
                             Published by Her Majesty's Stationery
                             Office London Col.1502

143. BRAYBROOKE, D and   A Strategy of Decision. Published the
     LINDBLOM, C E   Free Press, New York, 1970 Edition
                             p.81-110.

144. POPPER, K   The Poverty of Historicism. Published
                             Routledge & Kegan Paul London, Second
                             Edition 1969 reprint.

145. Cmnd 3638   The Civil Service Vol.1 Report of the
                             Committee chaired by Lord Fulton
                             Published by Her Majesty's Stationery
                             Office London 1968. p.17.

146. Cmnd 3638   The Civil Service Vol.1 Report of
                             the Committee chaired by Lord Fulton.
                             op cit.pp 61-62

147. CLARKE, Sir R   New Trends in Government. Published by
                             Her Majesty's Stationery Office London
                             p.81.

148. WOOTTON, G   Interest Groups op cit. pp 1-19.

149. MADGWICK, P J   Introduction to British Politics.
                             Published by Hutchinson Educational
                             Ltd. London 1970. p.339

150. ROSE, R   Politics in England. Published by
                             Faber and Faber London 1965. pp.126-
                             139.

151. FINER, S E   Anonymous Empire. Published by Pall
                             Mall Press London Second Edition 1966.
                             pp 1-5.

152. MOODIE, GC and   Opinion, Public and Pressure Groups
     STUDDERT-KENNEDY G   Published George Allen and Unwin Ltd.
                             London 1970. pp 59-62

153. ROBERTS, G K   Political Parties and Pressure Groups
                             in Britain. op cit. p.8.

154. ROBERTS, G K   Political Parties and Pressure Groups
                             in Britain. op cit. p.78

155. OLSON, M   The Logic of Collective Action.
                         Published Harvard University Press,
                         Cambridge, Massachusetts 1971. p.8

156. WOOTTON, G   Interest Groups op cit pp.30-44

157. MOODIE, G C and   Opinions, Public and Pressure Groups
     STUDDERT KENNEDY   op cit pp.63-64

158. HANSON, A H and   Governing Britain. Published Fontana/
     WALLES, M   Collins London 1971. pp.150-151.

159. DUVERGER, M   The Idea of Politics. Published
                         Methuen & Co.Ltd. London 1967.
                         pp. 116-120.

160. ROBERTS, G K   Political Parties and Pressure Groups
                         in Britain. op cit. pp. 89-96

161. OLSON, M   The Logic of Collective Action op cit
                         pp.49-51

162. BEER, S H   Modern British Politics. Published
                         Faber and Faber, London 1965. pp.319-
                         339.

163. STEWART, J D   British Pressure Groups. Published
                         Oxford University Press, Oxford 1958
                         pp.15-21

164. STEWART, J D   British Pressure Groups op cit
                         pp.152-204 and pp.250-265

165. STEWART, J D   British Pressure Groups op cit pp.244

166. ROBERTS, G K   Political Parties and Pressure Groups
                         in Britain. op cit pp.79-81

167. ROBERTS, G K   Political Parties and Pressure Groups
                         in Britain op cit pp.149-153

168. OLSON, M   The Logic of Collective Action op cit
                         pp.132-135.

169. OLSON, M   The Logic of Collective Action op cit
                         pp.141-148.

170. REYNOLDS, P A   An Introduction to International
                         Relations. Published Longman Group
                         Limited. 1971 p.179 and p.258.

171. Cmnd 5034   Report of the Committee on Safety and
                         Health at Work chaired by Lord Robens.
                         Published by Her Majesty's Stationery
                         Office 1972 pp.107-109

172. Cmnd 5034   Report of the Committee on Safety and
                             Health at Work op cit p.XV

173. Cmnd 5034   Report of the Committee on Safety and
                             Health at Work chaired by Lord Robens.
                             Volume 2 Selected Written Evidence.
                             Published by Her Majesty's Stationery
                             Office 1972 p. VII-IX.

174. Cmnd 5034   Report of the Committee on Safety and
                             Health at Work op cit p.1

175. Cmnd 5034   Report of the Committee on Safety and
                             Health at Work op cit. p.152

176. Cmnd 5034   Vol.2 of the Report of the Committee on
                             Safety and Health op cit. p. 1-5

177. Cmnd 5034   Vol.2 of the Report of the Committee on
                             Safety and Health op cit p. 7-14.

178. Cmnd 5034   Vol.2 of the Report of the Committee on
                             Safety and Health op cit p.56-63

179.   Labour's Programme. The Times 8 June
                             1973. p.4-5

180. Cmnd 5034   Vol.2 of the Report of the Committee on
                             Safety and Health op cit p. 111-148.

181. Cmnd 5034   Vol.2 of the Report of the Committee on
                             Safety and Health op cit. p.525-543

182. Cmnd 5034   Vol.2 of the Report of the Committee on
                             Safety and Health op cit p.599-606

183. Cmnd 5034   Vol.2 of the Report of the Committee on
                             Safety and Health op cit p.159-169.

In the following are given the composition of the Road Research
Laboratories Committees related to safety during 1969, and the
composition of the Royal Automobile Clubs Public Policy
Committee as it was in 1970, and also the RAC's Working Party
on Road Safety.

THE ROAD RESEARCH LABORATORY RESEARCH PROGRAMME REVIEW
COMMITTEE 1969

Terms of Reference

To review the programme of road research, and its applications,
and to consider its adequacy, deployment, orientation and
priorities in relation to available manpower and money.

Chairman

J.A. Jukes, Esq.C.B.	Ministry of Transport

Members

C.D. Foster,Esq.(until September 1969)	Ministry of Transport
D.J. Lyons,Esq. BSc, CEng, FRAeS, Hon.M.Inst.H.E.	Road Research Laboratory
J.R. Madge, Esq.	Ministry of Transport
R.S. Millard, Esq. CMG,BSc, PhD, CEng, FICE, M.I.Struct.E.,M.Inst.H.E.	Road Research Laboratory
H. Perring, Esq. MA, CEng. FIMechE, AMICE, AMIEE	Ministry of Transport
J.L. Paisley, Esq. MBE MEng MICE	Ministry of Transport
B.T. Price, Esq.	Ministry of Transport

THE ROAD RESEARCH LABORATORY RESEARCH COMMITTEE ON ROAD
SAFETY 1969.

Terms of Reference:

To advise the Director of Road Research on the planning, conduct
and application of research on road safety as it affects the
road user and the design of roads and road vehicles.

Chairman

H. Taylor, Esq., BSc, ACGI CEng, MIMechE.	Road Research Laboratory

Members:

B.N. Bebbington, Esq. OBE, MA	Home Office
F.J.S. Best, Esq., BSc(Eng) CEng, AMMunE,FICE,MIHE	Ministry of Transport
I.D. Brown, Esq. BSc, PhD	Medical Research Council
J.P. Bull, Esq., MA, MD, MRCP	Medical Research Council
P.J. Chapman, Esq. MB	Medical Research Council
T.Corlett,Esq., MA,FIS,AMIPA	J.Walter Thompson & Co.
Prof.J.R. Ellis, MSc(Eng) PhD, MIMechE	Advanced School of Automobile Eng.
Prof.W.F. Flloyd, BSc, PhD (Lond), FIP	Dept. of Ergonomics & Cybernetics Loughborough University of Technology
H.N. Jenner,Esq. MBE,MICE, MIMunE, DPInstHE	Hampshire County Council
K.J. Jones, Esq. BSc	Joseph Lucas (Electrical) Ltd.
Prof. H. Kay, MA, PhD	Dept. of Psychology University of Sheffield
Miss E.P. Kruse	Ministry of Transport
Prof.R.H.MacMillan, MA(Cantab) CEng, FIMEchE,MIEE,MSAE	Motor Industry Research Association
J.T. Manuel Esq.	Home Office
J. McGowan Est.	British Leyland Motor Corp.Ltd.
H. Perring, Esq. MA,CEng, FIMechE,AMICE, AMIEE	Ministry of Transport
K. Sargent, Esq.	Dept. Education and Science

W.M. Smith, Esq, MA   Aberdeen City Police

K.J.B. Teesdale, Esq.   Ford Motor Co.

Prof. W.D. Wright, DSc, DIC   Dept. Physics, Imperial College
                                 of Science and Technology.

THE ROYAL AUTOMOBILE CLUB'S PUBLIC POLICY COMMITTEE 1970

MR. LEONARD F DYER (Chairman)

    formerly Senior Vice-Chairman RAC
    President Traffic Commission of the Federation
      Internationale de L'Automobile (FIA)
    Vice-President, Traffice Commission of the World Touring
      and Automobile Organisation (OTA)
    Vice-Chairman, OTA/PIARC Joint Committee
    Member, Institute of Traffic Engineers of the USA
    Vice-President, British Automobile Racing Club

MR. WILFRID ANDREWS, CBE

    Chairman RAC
    President, Federation Internationale De L'Automobile (FIA)
    President, World Touring and Automobile Organisation (OTA)
    President, Commonwealth Motoring Conference
    President, Roads Campaign Council
    formerly a member of the National Road Safety Advisory
      Council
    a Steward of the RAC

THE HON. SIR CLIVE BOSSOM BART, MP

    MP for Leominster (Herefordshire)
    formerly Secretary, Conservative Transport Committee
    Member, Joint Committee of the Red Cross and St. John

MR. ARTHUR B. BOURNE CIMechE

    formerly Senior Vice-Chairman RAC
    Chairman RAC Associate Committee
    Chairman RAC Motor Cycle Committee
    formerly Director, Associated Ilifee Press Ltd.
    formerly Editorial Director, 'Autocar' 'Motor Cycle'
      'Motor Transport' 'Flight' and other technical
      publications.

THE MOST HON. THE MARQUESS CAMDEN DL, JP

    a Steward of the RAC
    formerly Senior Vice-Chairman RAC
    Chairman RAC Competitions Committee
    Member, Committee of the Order of the Road
    Chairman, Awarding Committee for the Segrave Trophy
    President, Auto-cycle Union.

THE RT HON LORD CHESHAM P.C.

>   Executive Vice-Chairman, RAC 1965-70
>   Chairman, British Road Federation
>   formerly Joint Parliamentary Secretary, Ministry of
>   Transport and Chairman, Departmental Committee on
>   Road Safety
>   formerly a Member of the National Road Safety Advisory
>   Council
>   Transport consultant to National Car Parks Ltd.

MR. NORMAN E DIXON, OBE

>   Chairman, Auto-Cycle Union
>   Deputy Chairman, Speedway Control Board
>   Chairman Tourist Trophy Race Committee of the Auto-Cycle
>   Union since 1947
>   Vice-President, Federation Internationale Motor-Cycliste

MR. ARTHUR D GILL

>   Member, Countryside Advisory Panel,Hampshire County
>   Council
>   Member,Petersfield Rural Council (Planning and other
>   committees)
>   Chairman, Petersfield Amenities Society
>   Member, Executive of the Hampshire Branch, National
>   Farmer's Union.

MR. H.N. GINNS BSc(Eng), FICE, FInstHE

>   Past President, Institution of Highway Engineers
>   formerly Deputy Chief Engineer, Ministry of Transport.

SIR WILLIAM H. GLANVILLE, CB, CBE, FICE, FRS

>   Engineering consultant
>   formerly Director of Road Research
>   Past President, Institution of Civil Engineers

THE RT HON THE EARL OF HALSBURY, FRS

>   a Steward of the RAC
>   formerly a Member of the Nationalised Transport Advisory
>   Council
>   formerly Managing Director, National Research Development
>   Corporation

THE RT HON THE EARL HOWE,CBE

>   President, British Automobile Racing Club
>   Chairman Buckinghamshire Road Accidents Prevention
>   Committee
>   Buckinghamshire representative on Highways Committee of
>   the County Councils Association
>   Member of ROSPA House Advisory Panel
>   President, Fiat Motor Club of Great Britain
>   Member, Ministry of Transport Control of Motor Rallies
>   Advisory Committee

MR. A H MATHIAS, CBE (Until his death in July)

    formerly Senior Vice-Chairman RAC
    Member, Industrial Court

MR. ROWLAND NICHOLAS, CBE, BSc(CEng), FICE, FIMunE, PPTPI

    Engineering and planning consultant
    formerly City Surveyor of Manchester
    Past President, Institution of Municipal Engineers
    Past President, Town Planning Institute

MR. HAROLD NOCKOLDS

    Deputy Chairman and Editorial Director, IPC Transport
      Press Limited
    formerly motoring correspondent of 'The Times'

THE RT. HON LORD NUGENT OF GUILDFORD, PC

    MP for Guildford (Surrey) 1950-66
    Parliamentary Secretary, Ministry of Agriculture and
      Fisheries 1951-57
    Joint Parliamentary Secretary, Ministry of Transport
      1957-59
    a Vice President, Royal Society for the Prevention of
      Accidents
    Chairman, Standing Conference on London and South East
      Regional Planning
    Chairman, Thames Conservancy Board
    President, National Association of River Authorities

SIR THOMAS PADMORE, GCB, KCB, CB, MInstT

    formerly Permanent Secretary, Ministry of Transport
    Member of Metrication Board and Chairman of its
      Steering Committee for the Transport Communication
      Industries.

THE RT. HON LORD REA OF ESKDALE, PC, OBE, DL

    Chief Liberal Whip, House of Lords 1950-55
    President of the Liberal Party 1955
    UK Delegate to Council of Europe, Strasbourg 1957
    formerly Chairman, Hire Purchase Committee of the
      Society of Motor Manufacturers and Traders

CAPTAIN SIR HENRY STUDDY, CBE, KPM

    formerly, in succession, Chief Constable of Northumberland
      Chief Constable of the County of Durham and Chief
      Constable of the West Riding of Yorkshire
    formerly represented County Chief Constables on the
      Departmental Committee on Road Safety
    Member, Road Research Board, 1959-63
    Head of RAC Public Policy Executive   MR A J A LEE

RAC WORKING PARTY ON ROAD SAFETY

MR. WILFRID ANDREWS (CHAIRMAN)	Chairman of the RAC
SIR CYRIL BIRTHCHNELL	Formerly a Deputy Secretary, Ministry of Transport
MR. ARTHUR BOURNE	Chairman of the RAC Motor Cycle Committee
DR. A. FOGG	Director of the Motor Industry Research Association
SIR RICHARD NUGENT, MP	Formerly Joint Parliamentary Secretary, Ministry of Transport and a Vice-President of the Royal Society for the Prevention of Accidents
MR. J.H. PITCHFORD	Chairman of the Road Research Board's Committee on Vehicles
SIR HENRY STUDDY	formerly Chief Constable of the West Riding of Yorkshire

The Farmer Safety Criterion was first postulated as the logical
basis for a method of assessing the safety of nuclear reactors
by Mr. F.R. Farmer, Director of the Safety and Reliability
Directorate of the United Kingdom Atomic Energy Authority, at
the IAEA Symposium on the "Containment and Siting of Nuclear
Power Reactors" held in Vienna, in April 1967. Although
originally only intended for the assessment of nuclear reactors
the criterion is so broad in concept that it can be applied to
the quantitative evaluation of the acceptability of any risk
situation.

In its very simplest form the criterion could be stated as
follows:- The acceptability of particular risks associated
with the activity of interest should be evaluated in
quantitative terms and the consequences of the whole spectrum
of risks compared with levels of risk that are known to be
generally acceptable. If the level of risk is higher than can
be accepted then the engineering of the activity must be
improved to bring the risk to an acceptable level.

From this statement of the criterion it can be seen that there
are two essential parts to it. First there are the risks of the
events occurring, and secondly there are the consequences of the
various events. It is fundamental to this form of evaluation
that the risks and consequences are evaluated in quantitative
terms, and that emotionally weighted judgements that particular
events are either credible or incredible are not used.

In a complex system that consists of a number of smaller systems
the probability of failure of the whole system can be built up
from the failure rates of the smaller systems. The analysis of
the failure of complex systems can be performed by using
computer programmes such as NOTED, described by Woodcock in
Nuclear Engineering and Design Vol.13 No.2

Similarly the consequences of each failure can be evaluated in
terms of some measurable quantity, it may be the number of
deaths that result from the failure; the cost in terms of lost
production, damage to the environment. Experience in applying
this type of analysis has shown clearly that if the analysis
cannot be performed because of lack of data then there is a
genuine uncertainty about the implications of the consequences
of a particular risk, and that further research is required
before an assessment of the risk can be completed.

In the case of a nuclear reactor the probability of various
accident conditions can be assessed against the resulting
release of I_{131} and the casualties it might cause to the
population living near it. Fig.4 shows a release probability
relationship proposed for reactors. If the analysis of a
particular system showed that it could lead to a release
probability above the target line some improvement to the system
would be required to make it acceptable. A system whose
release probability characteristic was below the line would be
acceptable. Taking the analysis a step further assuming
reactors have release probability equal to the target line and
are located in semi-urban sites then it has been shown by Bell
in Nuclear Engineering and Design Vol.13, No.2 that the
probability casualty relationship would be as shown in Fig.5

The generalised form of risk consequence analysis demanded by
the Farmer Criterion is shown in Fig.6. The target line would
vary according to the particular type of risk being considered,
although if the risk is being evaluated in terms of deaths
there is probably a universally acceptable target line.

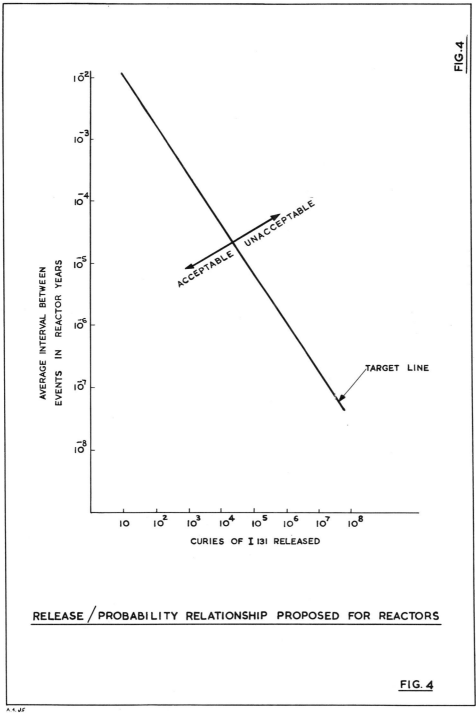

FIG.4

RELEASE / PROBABILITY RELATIONSHIP PROPOSED FOR REACTORS

FIG. 4

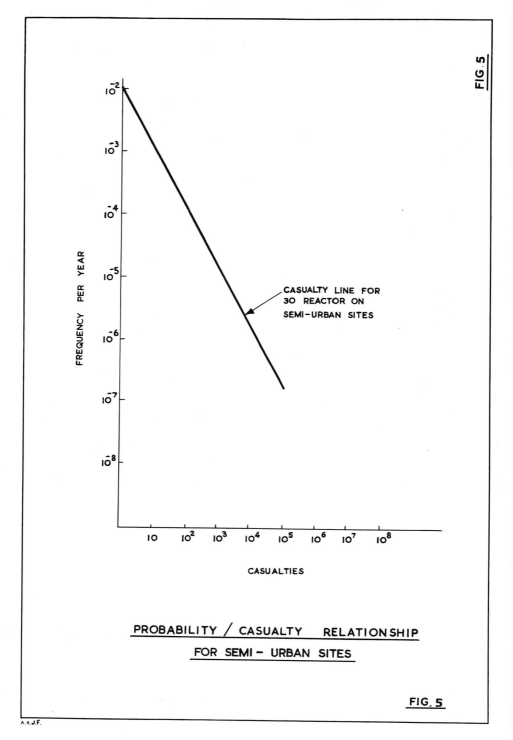

FIG. 5

CASUALTY LINE FOR
30 REACTOR ON
SEMI-URBAN SITES

FREQUENCY PER YEAR

CASUALTIES

PROBABILITY / CASUALTY RELATIONSHIP

FOR SEMI - URBAN SITES

FIG. 5

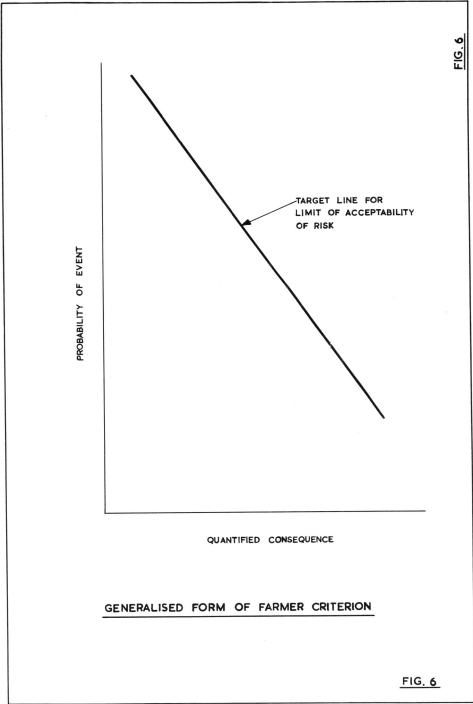

FIG. 6

PROBABILITY OF EVENT

TARGET LINE FOR
LIMIT OF ACCEPTABILITY
OF RISK

QUANTIFIED CONSEQUENCE

GENERALISED FORM OF FARMER CRITERION

FIG. 6

A.4.J.F

In reply to a Parliamentary Question on 22nd February 1968 Mr. Marsh gave the following details of the professional and business qualifications of the members of the Nuclear Safety Advisory Committee:-

Chairman

Sir Owen Saunders, FRS, MA,DSc   Professor of Mechanical Engineering, Imperial College of Science and Technology; Vice-Chancellor, University of London

Members

Mr. C.A. Adams, BSc,FInstP   Chief Nuclear Health and Safety Officer, Central Electricity Generating Board

Dr. T.E. Allibone, CBE, FRS, DSc, PhD(Cantab), PhD (Sheffield) MIEE, FInstP   Chief Scientist, Central Electricity Generating Board

Prof. A.L.L. Baker, DSc, Hon ACGI, MICE,MIStructE   Professor of Concrete Structure and Technology, University of London

Mr. G.F. Bullock, MA   General Manager, Vulcan Boiler and General Insurance Co.Ltd. Manchester

Mr. H. Cartwright, MBE, MA   Director of Water Reactors, Reactor Group, United Kingdom Atomic Energy Authority, Risley

Mr. W.J.C. Plumbe   H.M. Chief Inspector of Factories

Mr. Patrick Conner, OBE   Scottish Regional Officer (retired), Amalgamated Engineering Union

Prof. P.I. Dee, CBE, FRS MA(Cantab)   Professor of Natural Philosophy, University of Glasgow

Prof. J. Diamond, MSc MIMechE, BSc   Professor of Mechanical Engineering, University of Manchester.

Mr. P.T. Fletcher, CBE, BSc, MICE, MIMechE, MIEE   Managing Director GEC (Process Engineering) Limited

Mr. Trevor Griffiths, CBE,   Chief Inspector of Nuclear
  BSc, MICE,MIMechE, MIEE   Installations

Mr. F. Hayday, CBE   National Industrial Officer to
                                        the National Union of General
                                        and Municipal Workers, Vice-
                                        President of the TUC, and
                                        Vice-Chairman of its General
                                        Council 1963-64.

Mr. J.M. Kay, MA, PhD   Chief Engineer, Richard Thomas
  MIMechE, MIChemE   and Baldwins Ltd.

Dr. John F. Loutit, CBE, FRS   Director, Radiobiological
  MA, DM, FRCP   Research Unit, Medical Research
                                        Council, Harwell

Dr. A.S. McLean, MB, ChB,   Director, Health and Safety
  DIH   Branch, United Kingdom Atomic
                                        Energy Authority.

Colonel G.W. Raby, CBE,   Chairman and Managing Director,
  MIMechE, MIEE   Atomic Power Constructions Ltd.

Dr. R. Scott Russell, MA   Director, Radiobiological
  PhD, DSc   Laboratory, Agricultural
                                        Research Council, Wantage.

Mr. R.F. Jackson, MA(Cantab)   Director of Reactor Technology
  MIMechE, AMIEE   Reactor Group, United Kingdom
                                        Atomic Energy Authority, Risley

Secretaries

Mr. W.R. Loader and Mr. W.S. Gronow

Replacements are currently being invited to fill the seats of
the following members, recently deceased or retired:

Sir John Cockroft (Deceased)   Lately Master of the Churchill
  OM, KCB, CBE, FRS, MA,   College, Cambridge.
  PhD, MSc (Tech)

Dr. S.C. Curran, FRS, FRSE,   Principal, University of
  DSc, MA, BSc, PhD   Strathclyde

Mr. H.N. Pemberton (Deceased)   lately Chief Engineer Surveyor
  MIMechE   to Lloyd's Register

The following information about the membership of the Commission
is based on that given in Cmnd 4585, the first report of the
Royal Commission on Environmental Pollution and the 1971 edition
of Who's Who.

Sir Eric Ashby

Master of Clare College,
Cambridge. Fellow of the Royal
Society. Chairman Royal
Commission on Environmental
Pollution since 1970. 1942-46
Director Scientific Liaison
Bureau. 1951-59 Advisory Council
on Scientific and Industrial
Research. 1967-69 President of
British Association for the
Advancement of Science.

The Lord Bishop of Norwich

Educated at Cambridge and Yale.
Chaplain and Geologist British
Graham Land Expedition to the
Antarctic 1934-37. Director of
Scott Polar Research Institute
Cambridge 1947-49. Hon.Vice
President Royal Geographical
Society 1961.

Sir Solly Zuckerman, OM,KCB
Kt.FRS,MA,MD,DSc,MRCS,FRCP

Chief Scientific Advisor to HM
Government. Chairman Central
Advisory Council for Science and
Technology. 1960-64 Chairman
Committee on Scientific Manpower.
1948-64 Chief Scientific Advisor
to the Secretary of State for
Defence

Sir John Winnifrith, KCB

Permanent Secretary Ministry of
Agriculture, Fisheries and Foods
1959-67. Director General
National Trust 1968-70.

A.L.O. Buxton, ESQ. MC

Director of Anglia Television
since 1958. Extra Equerry to
Duke of Edinburgh 1964. Founder
of Stansted Wildlife Park.
Wildlife Film producer.

W. Beckerman, Esq.

Prof. of Political Economy, University of London since 1969. Head of Department of Political Economy, University College of London since 1969. 1967-69 the Economic Advisor to the Board of Trade.

F.F. Darling, Esq. Kt, SDc, PhD, LLD,FIBiol, FRSE

Vice President Conservation Federation, Washington DC. 1930-34 Chief Advisor Imperial Bureau of Animal Genetics. 1930-34 Leverhalm Research Fellow

N.A. Iliffe, Esq. CBE

Managing Director Shell Chemicals UK Ltd. since 1960. Deputy Chairman since 1968, also Chairman Petrochemicals Ltd. and Subsidiaries 1969. Hon. Fellow University of Manchester Institute of Science and Technology 1969.

V.C. Wynne-Edwards, Esq. FRD, MA, DSC, FRSC, FRSE

Regius Prof. of Natural History, University of Aberdeen since 1946.

136

INTRODUCTION

Preliminary investigations had shown that in the open
literature there was apparently no significant body of data
about the organisation and role of interest groups concerned
with influencing policy on hazard control. A survey was
designed to overcome this shortage and to provide a uniform body
of data on which to base analysis of the resources and interactions interest groups appear to have had with the policy
making process. The survey took the form of a questionnaire
circulated to a selected sample of interest groups. In some
cases the organisations invited me to discuss their answers to
the questionnaire, and these interviews provided additional
information.

THE DESIGN OF THE QUESTIONNAIRE

To determine what kind of response could be expected to the
questions included in the questionnaire I wrote to the RAC on a
personal basis and put to them some of the questions I hoped to
have answered. They answered the questions in full and
amplified their answers with a considerable amount of
documentation. Thus encouraged, the design of the final
questionnaire was completed.

The design had to take into account the type of organisation to
be surveyed, which had to some extent been selected by the early
parts of the study. The organisations selected, which could be
regarded as a modified stratified sample of the interest group
population, consisted of representatives of the three main types
of interest group, i.e. economic, integrative and cultural, and
were mainly representative of what Wootton defines as the third
order level of operation. A particularly important feature of
the groups as far as the design of the questionnaire was
concerned, was that with one exception they all had some full-
time staff, so it was possible to assume that the questionnaire
would be dealt with by reasonably literate people.

With this sample framework it was possible to aim the design of
the questionnaire at obtaining quantitative data about the
resources and organisation of interest groups.

A copy of a completed version of the questionnaire is given at
the end of this appendix.

The purpose of each of the questions was as follows:-

The first question was simply a matter of identifying the organisation and needs no further justification.

Questions 2 to 4 were aimed at identifying the resources of the organisation. Question 2 attempts to do this by asking how many members and employees an organisation has. The number of staff the organisation has gives an indication of effort available to prepare papers, to undertake research, and to present the organisation's case.

Question 3 was aimed at establishing whether the organisation represented a particular section of the community such as industry or professional engineers, or whether it represented a cross section of the community as a whole.

The fourth question, which was about the financial resources of the organisation, was intended to determine if, because the organisation owed its financial well being to a particular section of the community, it would be biased in its views, the implication being that the greatest source of financial support would have the greatest influence over the policy of the organisation.

Questions 5 and 6 were designed with the object of determining the goals of the organisation by establishing how the organisations formed their policy and what their view was of the form government policy should take on hazard control. The answer to question 5, it was hoped, would reveal whether the formation of the organisation's policy was mainly the product of the full-time executive or the product of some procedure that allowed direct participation of the general membership.

Question 7 was directed at determining how far the organisation understood and accepted the probability type of argument that is currently used in the aircraft and nuclear industries to evaluate the acceptable level of hazards. It was hoped that this question would indicate the depth of technical understanding that the organisation had of hazard control problems.

Questions 8, 9 and 10 were designed to explore the nature of the interaction of the organisation with government bodies and committees, in other words their relationship with proximate policy makers. It was hoped that question 9 would show what combination of formal and informal links existed. The reason for asking in question 10 if a public relations specialist was employed was to try and determine the effort that the organisation was prepared to put into getting its view accepted.

Questions 11 and 12 were designed to determine the role that these organisations had played in the past in influencing government hazard control policy.

When the questionnaire was designed it was appreciated that perhaps the answers to some of the questions would be contained in documents already prepared by the organisations and they might prefer to illustrate their answers by sending copies of such reports. This was the reason for question 13.

Question 14 was simply designed to establish if there was likely to be any restriction on the publication of the information presented.

Response to Questionnaire

Thirty-six questionnaires were sent out. The list of organisations to which they were sent is given in Table 2. Twenty-seven of the organisations replied in some form, giving a 75% response, which compares very favourably with the average return of 20% for postal enquiries mentioned by Ilersic (1). Four of the organisations, The Society of Motor Manufacturers and Traders, The Confederation of British Industry, The British Medical Association and the Association of Municipal Corporations, gave interviews to allow them to amplify the answers they wished to give to the questionnaire.

TABLE 2

LIST OF ORGANISATIONS THE QUESTIONNAIRE WAS SENT TO

British Roads Federation	
Society of Motor Manufacturers and Traders	R
Motor Industry Research Association	R
Institution of Civil Engineers	R
Road Operators Safety Council	
Institution of Mechanical Engineers	R
Royal Society for the Prevention of Accidents	R
Automobile Association	R
Royal Automobile Club	R
Centre for Study of Responsive Law	
Motoring Which (Consumers Association)	R
Society of British Aerospace Companies	R
Lloyds Aviation Underwriters Association	R
Royal Aeronautical Society	R
Guild of Air Traffic Control Officers	R
British Air Line Pilots Association	R
Air Registration Board	R
Flight Safety Committee	R
Amalgamated Union of Engineering Workers	R
Central Electricity Generating Board	R
Nuclear Plant Contractors (TNPG and BNDC)	
Institution of Professional Civil Servants	R
Confederation of British Industry	R
National Union of General and Municipal Workers	R
British Medical Association	R
Institution of Public Health Inspectors	
Association of Public Health Inspectors	R
National Society for Clean Air	R

TABLE 2 (Continued)

Clean Air Council	R
Civic Trust	R
Urban District Councils Association	
Association of Municipal Corporations	R
Conservative and Unionist Central Office	R
Labour Party	
Liberal Party	

36 Questionnaires were sent out

The 27 organisations that replied in some form are marked with an R.

In sending the questionnaires out care was taken to address them to the Chief Executive of the particular organisation, and in fact, in most cases the Chief Executive appear to have answered the questionnaires themselves, so the answers given genuinely reflect the policy of the organisation as seen by the Chief Executive. Three of the 4 interviews were also conducted by chief permanent officials of the organisations.

There appears to be no certain way of accounting for the fact that some organisations did not answer the questionnaire. Of the 11 organisations associated with road transport, 8 replied. All 7 of the organisations associated with air transport responded, although with one organisation only a part reply was obtained after protracted correspondence. All 4 organisations associated with factory hazards responded, but only 2 of the 5 organisations associated with the nuclear industry responded. Seven of the 10 questionnaires sent to organisations dealing with air pollution were returned fully completed. Only one political party replied.

Summary of Data Collected

The data collected by the questionnaires are summarised in the following, under the five activity headings. In some cases the answers to the questionnaire are supplemented by information gleaned from other sources.

Road Transport

The response to the questionnaire was quite satisfactory in the road transport area, eight of the eleven organisations responded in full. Answers were obtained from the Society of Motor Manufacturers and Traders, the Motor Industry Research Association, the Institution of Civil Engineers, the Institution of Mechanical Engineers, the Royal Society for the Prevention of Accidents, the Automobile Association, the Royal Automobile Club and Motoring Which (the Consumers Association).

The Society of Motor Manufacturers and Traders

The Society of Motor Manufacturers and Traders gave me an interview at which they discussed in detail the answers to the questionnaire. The Society has been in existence since 1902, and it perceives one of its main functions as providing the motor industry with a means of formulating, making known and influencing general policy affecting the industry. Membership of the Society offers the right to display products at the Society's shows, access to the Society's advisory service on technical, marketing and legal questions. There are about 1,500 members of the Society and they include vehicle manufacturers, component manufacturers and major distributors. These members pay a subscription related to turnover, and the Society has an annual budget of about three quarters of a million pounds. The Society has a staff of 120 including a Technical Department of 18.

The governing body of the Society is the Council which consists of approximately 100 annually appointed members. Responsible directly to the Council are the Executive Committee, The General Purposes Committee, the Roads and Transport Policy Committee, and the fourteen section committees which look after the sectional interests of the industry such as cars, commercial vehicles, accessories, tyres, etc. Committees dealing with external trade, public affairs, economics, and technical matters are all responsible to the Executive Committee. These committees deal with specific problems affecting the industry as they arise.

The view of the Society on the way government policy should develop on the control of hazards associated with the motor industry was that if type approval of cars was to be adopted this should take the form of the EEC approvals rather than some form of national certification.

In reply to the question about the acceptability of using probability techniques to determine whether or not a particular hazard could be sanctioned the opinion was expressed that in some situations the probability arguments were suitable, but the level of probability that was acceptable was a matter for the companies to investigate.

The contacts that the SMMT has with the government are at both the formal and informal level. At the formal level SMMT is represented on various government working parties and it has regular meetings with the Department of the Environment. Staff are in day to day working contact with their opposite numbers in the civil service. If SMMT consider a particular subject is of sufficient importance they would make representations direct to the Minister concerned. Informal contacts develop through membership of the British Standard Institution, the Royal Society for Prevention of Accidents and the Motor Industry Research Association.

SMMT does have a public relations department, which is responsible for presenting the views of the motor industry to the public.

The Society had been successful in the past on a number of occasions in having the timing of the introduction of legislation adjusted to take account of the time manufacturers require to modify their designs and production to take account of the requirements introduced. As a result of these representations it is now an established procedure for new regulations to allow a satisfactory lead time to enable manufacturers to carry out the necessary re-design and modifications of production schedules.

From the above, it can be seen that the SMMT is an interest group with extensive financial and technical resources. Its goals are to further the interests of the motor industry, and decision makers will see it as representing the views of the motor industry. The formal and informal contacts that the Society has with government are extensive and influential and in the Society's view have been of mutual benefit.

The Motor Industry Research Association.

The Motor Industry Research Association's answers to the questionnaire were such that a full evaluation of its influence on policy could not be made. The Association has two hundred members, who are all British companies, and financially support the Association by subscriptions fixed in proportion to the companies' turnover. The Association employs a staff of 200. Policy of the Association is defined by a Council made up of representatives elected from the membership.

The view expressed on the way government policy should be developed on the control of hazards was that it should be developed in consultation and in co-operation with industry.

Probability was not considered to be an entirely acceptable method of determining if a particular hazard can be sanctioned.

Contacts with government are essentially informal through personal contacts with the Department of the Environment, membership of British Standards Institution and the International Standards Organisations, and membership of organisations such as the Noise Council.

No indication was given of cases in which the Association had influenced government policy. Although from the fact that the Association through its research is a major source of technical information, it seems reasonable to assume that weight will have been given to opinions expressed by representatives of the Association. In the future it is possible that consideration of commercial security may inhibit the information that the Association's representatives may present, as from 1975 all the research that the Association undertakes will be paid for directly by the members on an individual contract basis, rather than collectively as at present.

The primary goal of the Association is to satisfy the research requirements of the motor industry. It has contacts with government, but may be inhibited from exploiting its influence by consideration of its commercial relationships with the manufacturers.

The Institution of Civil Engineers

The Institution of Civil Engineers, the senior British professional engineering society, answered the questionnaire fully. The Institution was founded in 1818, and its first President was the famous engineer Thomas Telford. In 1828 the Institution received its first Royal Charter. The object of the Institution, as expressed in the first Royal Charter (2) is, "promoting the acquisition of that species of knowledge which constitutes the Profession of Civil Engineer, being the art of directing the great sources of power in nature for the use and convenience of man". This objective is partly satisfied by the role the Institution plays in ensuring that Civil Engineers are properly trained and qualified.

The Institution has 43,500 members all of whom are private individuals. Twenty-eight thousand five hundred of the members are fully qualified corporate members. The income of the Institution is derived from the subscriptions of members and sale of publications. For the year ending 31 December 1971 the income account (3) shows that the total income of the Institution was £506,000.

The policy of the Institution is made by the general meeting of corporate members, and by the Council acting generally on the recommendations from the Committees. The committee structure of the Institution is extensive and beside the committees necessary for running the Institution there are specialist committees concerned with technical aspects of civil engineering international technical co-operation, education and training, and national and economic affairs. There is also a specialist group dealing with transportation engineering. The technical committees deal with subjects such as conditions of contract, piling, safety in civil engineering, safety of reservoirs, and the engineer and building legislation. The International Committee that the Institution is associated with includes the World Federation of Engineering Organisations, the International Association of Hydraulic Research, the International Commission on Large Dams, and the Permanent International Association of Road Congresses. Under the national and economic affairs heading the Institution runs a National Affairs Committee; it collaborated with the Buchanan Report Standing Committee, and participates with the Presidents of other professional institutions in the President's Committee on the Urban Environment. Also it has set up an ad hoc committee on human habitat/natural resources to present views on the human habitat to Lady Dartmouth's Working Party, which prepared a report for the UN Conference on the

Environment in Stockholm in 1972. Another ad hoc committee was
set up to prepare the evidence that Professor Lord Zuckerman
asked the Institution to submit to his Commission on Mining
and the Environment.

The range of interests of the Institution was indicated by the
major topics that the council discuss at their meetings, during
1971 the following topics were discussed:- The Industrial
Relations Bill, university intake for civil engineering courses,
Professional Interview policy, the Royal Commission on
Environmental Pollution, publications policy, Codes of Practice,
the future of research organisations and The Common Market.
The transportation engineering group held seven informal
discussions during 1971 which dealt with subjects such as
transportation in Paris and Stevenage, planning for leisure
travel, training of traffic and transportation technicians, and
the social cost-benefit studies for railways.

The view the Institution had on the way government policy should
be developed on the control of hazards was expressed as being:
"While the ultimate responsibility for public health and safety
must lie with government, the policy must acknowledge and rely
on professional advice, not stifle technical advances".

The opinion that the Institution expressed on the use of
probability techniques to evaluate the acceptability of a
particular hazard was that "in some fields probability is
considered to be an appropriate and unavoidable criterion
(e.g. design of flood-control works)"

The Institution is represented on about fifty government and
national committees that could be regarded as policy forming
centres and organisations that put them in close contact with
proximate policy makers. These committees include the Urban
Motorways Committee, the Parliamentary and Scientific Committee,
and the Department of the Environment Standing Consultative
Committee for Civil Engineering. Informal contacts, the
Institution claims, have developed through direct contact with
Ministers and Departments and through the meetings and
conferences that it organises. There is an Information Officer
on the staff of the Institution, and his duty is to disseminate
the knowledge and opinions of the Institution to interested
parties.

The Institution's influence is extensive as exercised in the
collective way, and as exercised in the individual way through
individual members of the Institution acting in their
professional capacity. The following is a list of the actions
the Institution has taken which it considers have influenced
government policy in relation to hazard control. The list does
not contain any specific reference to policy on roads, which
suggests that there was no significant comment on policy and
that the Institution's contribution was in this case related to
technical detail of road construction, as for example discussed
by the transportation engineering group.

LIST OF ACTIONS THAT THE INSTITUTION HAS TAKEN
AND THAT IT CONSIDERS HAVE INFLUENCED POLICY

(i) Interim report of the Committee on floods in relation to reservoir practice, 1933, with additional data on floods recorded in the British Isles between 1932 and 1957 - published 1960.

(ii) Reservoir safety. Report of the ad hoc Committee set up by the Institution to submit proposals for a revision of the Reservoirs (Safety Provisions) Act 1930. - published 1966

(iii) Flood studies for the United Kingdom. Report of the Committee on Floods in the United Kingdom - published 1967

(iv) Report on safety in civil engineering - published 1969

(v) Safety on construction sites. Proceedings of the Conference held March 1969 - published 1969

(vi) Hazards on construction. Conference held November 1971 - published 1972

(vii) Reports to the DoE, in preparation for the United Nations Conference on the Environment at Stockholm in June 1972

(viii) Evidence presented to Lord Zuckerman's Commission on Mining and the Environment

The Institution of Civil Engineers has extensive financial and technical resources, it has a very well established position in the government consultation machinery, and has made considerable efforts to satisfy all the obligations placed on it. This aspect of the Institution's work is in keeping with the goal set for the Institution in 1828, as promoting the knowledge in the art of directing the great sources of power in nature for the use and convenience of man. Many Members of the Council of the Institution are currently (1972) at or are very close to the proximate policy maker level. Relevant to the specific area of road transport they include County Surveyors, the Chief Highway Engineer for the Department of the Environment, partners in firms of Consulting Engineers, and leading engineers in the academic world. It is possible that these people are influenced by the discussions that take place in the Institution and that the exchange of views that takes place in the meetings of the transportation engineering group have some influence.

The Institution of Mechanical Engineers

The Institution of Mechanical Engineers is very similar to the Institution of Civil Engineers except that it is younger, founded in 1847 and it did not receive its Royal Charter until 1930 over 100 years later than the Institution of Civil Engineers. The Mechanical Engineer is concerned with the slightly narrower field of manufacture and operation of moving machinery rather than the broad interest of the Civil Engineer which is making use of all the sources of power in nature. It

is suggested by Parsons (4) that one reason for the founding of
the Institution of Mechanical Engineers might have been that
the Council of the Institution of Civil Engineers had refused
to admit the railway engineer George Stephenson as a member
unless he submitted a probationary essay as proof of his
capacity as an engineer. George Stephenson did, in fact,
become the first President of the Institution of Mechanical
Engineers. Initially the Institution of Mechanical Engineers
appears to have given precedence to people whose engineering
training was based more on practical experience rather than
extensive knowledge of the underlying theory, and it was not
until 1913 that entrance examinations were introduced. Although
now there is little significant difference in the academic
requirements for membership of any of the Institutions like the
Civil and Mechanical Engineers that are members of the Council
of Engineering Institutions. The membership of the Institution
of Mechanical Engineers is 73,500 of which 47,800 are fully
qualified Corporate members. For 1971 the total income of the
Institution was £501,000 so although the membership of the
Mechanicals is larger than the Civils, the income is about the
same.

The policy of the Institution is made by its Council elected
from the members, and its standing committees. No collective
view on the way government policy should develop on the control
of hazard was expressed, but it was suggested that professional
viewpoints are established on the technical aspects of hazards
and safety through the medium of conferences and other meetings.

With regard to the Institution's view of the acceptability of
hazards being evaluated in terms of probability, it was
suggested that the method might be acceptable in some cases but
not others.

The major difference between the Institution of Civil Engineers
and the Institution of Mechanical Engineers is that the Civils
are represented on government bodies whereas the Mechanicals
subscribe to the thesis that the Council of Engineering
Institutions is the spokesman for the professional engineering
bodies and is therefore not represented independently on any
government committees.

The Institution does not claim to influence government policy
in its role as an independent body, but it would be wrong to
maintain that the government does not observe and perhaps
include in their deliberations the technical observations which
emerge through the meetings and publications of the
Institution. At present the Institution does not employ a
public relations expert to promote its views.

Looking at the meetings held by the Mechanicals during the
1971-72 session there were three meetings which might have had
some impact on road transport safety thinking of policy makers
as follows:

1.   Presentation of a paper, "The primary safety -
        Vehicle design to avoid road accidents" by
        R.H. MacMillan

2.   A symposium on air pollution control in transport
        engines.

3.   Presentation of a paper "The conflict between
        traction and stability in passenger cars"
        by S.H. Grylls

Although the Institution of Mechanical Engineers has
considerable financial and technical resources it does not
perceive its role as that of having to attempt to influence
government policy. * The only influence it exerts comes about
indirectly through the meetings, conferences, and symposia
that it holds, which let people discuss technical problems and
develop solutions to them by discussion and the relatively
free exchange of ideas.

The Royal Society for the Prevention of Accidents

The Royal Society for the Prevention of Accidents which is a
company limited by guarantee and not having a share capital,
started life in 1916 as the London Safety First Council. The
Society's main purpose is to educate, inform and to alert the
public, industry, government and press, on the need for accident
prevention. At present it has 5,200 members, and 240 employees.
The members are mainly corporate bodies but there are some
private individual members such as industrial safety consultants
and driving instructors. The subscriptions that members pay is
related to the type of organisation they are, their size, and
the services of the Society that they use.

The President of the Society was Lord Beeching in 1970-71, nine
of the fiteen Vice-Presidents were members of the House of
Lords, three were privy Councillors and one was a Member of
Parliament. Lord Robens was one of the Vice-Presidents. The
Society is governed by an Executive Committee which is elected
annually. The policy of the Society is formed by the Executive
Committee on the basis of the advice it gets from the Management
and Finance Committee and the seven specialist committees.
These specialist committees are as follows:- National Road
Safety Committee, National Industrial Safety Committee, National
Safety Education Committee, National Water Safety Committee,
National Agricultural Safety Committee and a National Publicity
Committee.

No general opinion was expressed by the Society on the way
government policy should be developed on the control of hazards.
But it was stated that the Society's views on this question had
been expressed to the government and to the Robens' Committee.

* See note page 90

In reply to the question about the use of probability techniques
to evaluate the acceptability of particular hazards the Society
gave the following answer:- "Foreseeability is a more
appropriate word than "probability" since it embraces the need
to detect, recognise or assess probability. We are not sure
what you mean by "acceptability" but we presume you mean
acceptability in the sense of "no action". This issue rests on
the potential of the hazard (if it can be determined), i.e.
whether it can kill, maim, cause disaster, or at the maximum
cause minimal harm. Debate on this is complicated by the
laymen's confusion between "accident" and "injury". This answer
suggests little experience in the application of probability
techniques, as the first step in employing this technique is to
decide what the criteria for acceptability is, and you cannot
apply the technique without a very detailed knowledge of the
hazard and assessing probability. It is because the technique
demands the discipline of assessment that it is attractive.

In reply to the question about which government bodies and
committees related to the control of hazards the Society is
represented on, they replied:- "The Industrial Safety Advisory
Council (Department of Employment) and we work closely with all
Ministries concerned in any way with accident prevention and
certain outside bodies such as the British Standards Institute
and the Advisory Council for Education". The contact between
the government and the Society is considered to be a constant
dialogue, and is supplemented by government observers on the
Society's main committees.

The Society has an Information Services Department which deals
with press relations, publication of the Society's periodicals
and relations with the mass media generally.

The Society sent a delegation to the Minister of Transport
during 1970 to stress the importance of giving Road Safety
Officers the right status, training and responsibilities.
One hundred and four of the Society's staff are employed
specifically on road safety work, and 77 of these are paid for
by a Government grant from the Department of the Environment.
Much of the effort of this staff has been directed to educating
road users in more skilful and safe use of the roads.

The Society did not claim that it was solely responsible for
modification of policy, but that modification had been arrived
at in collaboration with other interested parties such as
manufacturers, trade associations and unions.

The Society has adequate financial resources for its purposes
and some technical resources, it has developed formal and
informal contacts with the government, and its interests are
well represented in the House of Lords. Particular emphasis is
given by the Society to education in road safety, such as by
improved training of road users. It is perhaps questionable
that the technical resources of the Society are adequate to deal
with all the technical problems associated with the wide range

of the Society's interest.

The Automobile Association

The Automobile Association had a membership, mainly private individuals, of 4,496,900 at the end of December 1971. It can therefore claim to represent a very large proportion of road users. It has a staff of approximately 5,750 and is supported mainly by annual subscription.

The policy of the Automobile Association is formulated by a committee on the advice of specialist staff, and after having taken into account the views made known to them of individual members. The AA view on the way government policy should be developed on the control of hazards was expressed as being the identification, evaluation and implementation of relevant remedial measures. This was a rather general statement but appears to indicate recognition that hazard control policy is likely to develop only in an incremental way.

The AA accepted without reservation the proposition that the acceptability of a particular hazard can be determined in terms of probability.

The AA is represented on a number of official committees and working parties relevant to motoring matters. It implements this formal contact with Government by direct negotiation with the appropriate authorities and on occasion in conjunction with other interested organisations. The AA has its own Press and Information Department to help promote its views.

The following six cases were quoted as being those in which the AA had attempted to lead or influence government policy in relation to hazard control:-

1. Opposition to 'random' breath tests

2. Advocacy of pre-driver training

3. Opposition to overall speed limit of 70 mph

4. Advocacy of simplification of rules concerning the parking of motor vehicles at night without lights

5. Advocacy of installing a modern emergency warning system on motorways

6. Advocacy of entirely radical system of new road construction finance including the setting up of a National Roads Authority

In cases 1, 2, 4 and 5, the AA considered they had government policy modified to take account of their views. The proposal for random breath tests has been dropped. Pre-driver training is now fairly widely accepted as part of school curricula. New rules for parking without lights have been introduced recently. Emergency warning systems are now (1972) being installed on motorways.

The resources the Automobile Association have are extensive both
in financial and technical terms, and these are reinforced by
the AA representing about 4½ million motorists. In other words
this latter resource represents a large number of voters that
no political party can afford to ignore completely. The
Association has well developed formal and informal contacts
with government, and claims positive success in getting its
views accepted. Like other organisations, its approach is
incremental, in that it tackles the problems as they arise.

The Royal Automobile Club

The Royal Automobile Club is very similar to the Automobile
Association, although it has a much smaller membership. The
RAC does not as a matter of practice give precise membership
figures but pointed out that membership has often been stated
in the press to be in the region of 1½ million. The Club's
Public Policy Committee (described in Appendix 1) is responsible
for forming the Club's policy in relation to legislation.
During 1970, this committee met five times. Mr. D.J. Lyons,
Director of the Road Research Laboratory, was invited to attend
a special meeting arranged to discuss new developments in
connection with road safety and traffic engineering. In many
matters of policy the Club acts with other motoring
organisations through the medium of the Standing Joint Committee
of the Royal Automobile Club, the Automobile Club, and the
Royal Scottish Automobile Club.

The Club takes pains to provide the press and Members of
Parliament with detailed information to ensure that the Club's
views are known on matters of motoring interest when they are
subject to debates or questions in Parliament.

There is no single statement of the Club's views on the way
government policy on the control of hazards should develop, but
the description given later shows the range of the Club's
concern.

On the question of the acceptability of evaluating a particular
hazard in terms of probability, the Club's view was that it was
a more expensive method than the motor transport industry could
afford, and that determination of hazards is generally evaluated
on the basis of the experience of motor and traffic engineers.
I think this view might prove hard to justify, particularly in
view of the AA's unequivocal acceptance of the proposition.

A list of the government committees and bodies the Club was
represented on was not provided. However, the Club has provided
assistance to the All-Party Roads Study Group at Westminster
and to the production of the Morgan Report - which expounded
the economic case for expansion of the road programmes. The
Club considered that in preparing the White Paper "Roads for
the Future - the New Inter-Urban Plan for England", the
Ministry of Transport had taken some account of their views
expressed on the earlier Green Paper and increased the mileage
of road to be built, although the targets were still considered

too low. The Club made representations whenever schemes were
prepared to give public transport vehicles priority, which it
was considered were likely to affect the interests of the
private motorist unreasonably. Such representations were made
at Southampton and\to the Greater London Council. The Club has
urged the government to speed up plans for a code of practice
to ensure that lorry loads are properly secured. The Club made
representations to the Department of the Environment to prevent
the placing of builders' skips on the highway which could cause
obstruction and danger. The necessary powers were included in
the Highway Act 1971, and the Department of the Environment
issued appropriate advice to local authorities. This means that
builders are now prohibited from leaving their skips on the
highway unless specific permission has been given by the
highway authority. When the proposals for reducing the number
of vehicle testing stations from 22,000 to 2,000 were put
forward, the Club was concerned to protect the interest of the
private motorist so that the changes to the system should not
cause unreasonable additional expense or inconvenience.

The Club also encouraged members to become safer motorists by
the publicity it gives to road hazards and regulations, and by
arranging and improving training facilities-

The Club's view on the role the government should play in
improving road safety is perhaps given in the last two
sentences of its working party on Road Safety report in 1964,
which states "It is essential that the government should make
greater efforts to secure the co-operation of the motoring
public. Any stricter enforcement measures must be accompanied
by appropriate constructive action, and the removal of
unreasonable or unnecessary restrictions". This description
does agree with the kind of views the Club expressed in the
above cases.

The Royal Automobile Club is very much the same kind of interest
group as the Automobile Association, it has considerable
resources and well developed formal and informal contacts.
The main difference between the two stems from the fact that the
Royal Automobile Club has the smaller membership and
consequently smaller resources.

MOTORING WHICH (CONSUMERS ASSOCIATION)

The research that the Consumers' Association undertakes does
have some impact on motoring safety policy. The Consumers'
Association has a membership of about 600,000 and a total
annual income of £2,000,000. All the members are private
individuals. The Association has 300 employees, and 25 of
these are concerned with "Motoring Which", the publication in
which the Association reports the results of its motoring
investigations. The policy of the Association is decided by a
Council of independent unpaid members. The Council reaches its
decisions after considering recommendations from the Director.
In order that the implications of the Association researches are
followed up in Westminster and Whitehall the Association has a

Consumer Campaign Committee made up of the Association's senior management, and this Committee ensures that the representations the Association makes are based on well researched information and are fully documented.

The Association's view on the way government policy should be developed on the control of hazards is that there should be a government department set up to deal specifically with Consumer affairs.

The Association had no definite opinion on the acceptability of a particular hazard being determined in terms of probability.

The Association is represented on the Executive Board of the British Standard Institution and three dozen technical committees of BSI. The Association is seeking representation through the International Organisation of Consumers Unions on the ISO Automobile Safety Committee.

The Association attempts to influence government policy on hazard control by direct representation to government departments, briefing and technical help for members of parliament, publication of their factual findings and conclusions in their own magazines; promotion of their views in the local and national press and broadcasting. To help in the promotion of the Association's views they employ public relations experts.

The Association claims to have been active in practically all recent cases where government policy in relation to consumer hazards has been redirected as a result of non-government influence. The view of the chairman of the Association in relation to car safety was expressed in her last annual report in the following terms: "While progress has at long last been made in formulating basic safety standards for most electrical appliances, stricter safety regulations are still urgently required governing standards of construction, design and performance of cars. All these reforms for whose need we have provided abundant evidence and will continue to press vigorously"

Many other organisations attempt to influence policy on road transport hazard control. Two that are discussed under other headings in this study, are the Civic Trust, and the Association of Municipal Corporations. The Civic Trust presented a report to the Minister of Transport in which it argued against an increase in the maximum permitted weight and size of road goods vehicles. The Association of Municipal Corporations has been concerned to influence policy on the winter maintenance of roads, creation of walkways, improving the maintenance of traffic signals, limiting the size of goods vehicles, the hours drivers of goods and passenger vehicles may drive and control of street lighting schemes. The Association also consider that the Highways Bill had given effect to the representations they had on the use of builders' skips in urban areas, and the use of scaffolding and tower cranes.

Air Transport

The survey made of interest groups active in the air transport
field of activity was the most successful of the five studied.
All seven organisations contacted responded in some way, and
four completed the questionnaire fully. The four organisations
that responded fully were The Society of British Aerospace
Companies Limited, The Air Registration Board, The British Air
Line Pilots Association, and the Guild of Air Traffic Control
Officers. The Flight Safety Committee and Lloyd's Aviation
Underwriters' Association gave sufficient information for an
assessment to be made of the extent to which they may influence
policy on hazard control. The Royal Aeronautical Society
declined to complete the questionnaire as the Secretary did not
consider the Society was actively interested in influencing
policy related to the control of hazards. However, from the
Annual Report, the Charter of Incorporation, and the By-Laws
of the Society, it is possible to obtain some indication of
the influence that it could have on policy formation.

The Society of British Aerospace Companies Limited

From the information Sir Richard Smeeton, Director of the
Society of British Aerospace Companies Limited, gave in answer
to the questionnaire it appears that the Society has 347 members
and 43 employees. All the members are corporate bodies, i.e.
air frame, aeroengine, equipment, and materials manufacturing
companies. The members are divided into two classes, ordinary
and associate. The ordinary members pay an annual subscriptio-
of £3,000 and a levy based on aerospace turnover, and associate
members pay annual subscriptions of between £125 and £500
depending on turnover. The Society Council is the policy making
body and is supported by The Equipment Group Committee, The
Materials Group Committee, The Contracts Advisory Committee,
The Finance Committee, The Sales and Export Committee, The
Taxation Committee, The Flight Operations Committee, The
Technical Board (which has a standing Committee on Airworthiness)
The Guided Weapons and Space Policy Committee, The Education
Committee and the Production Committee. The Flight Operations
Committee and the Technical Board are the two committees most
concerned with matters that could be related to hazard control,
The Council to which all the committees report consists of 26
members nominated by the ordinary members and associate 6 members
nominated by member Companies. Clearly the Society of British
Aerospace Companies is financed and controlled solely by the
companies involved in aircraft design and construction.

The Society expressed their view on the way government policy
should be developed on the control of hazards in the following
terms: "The Society's prime object is to encourage, promote
and protect the British Aerospace Industry and within this,
problems of airworthiness in manufacture and measures to promote
safety in the air, in the limited aspect of flying of
development aircraft, are major matters of concern. By its
representation on government policy committees the Society

plays its part in shaping the general development of hazard control".

The Society's opinion on the acceptability of hazard evaluation in terms of probability was that: "Modern airworthiness concepts are directly based on a national assessment of public tolerance of accidents. The standards of safety and reliability of vital equipment components and structural elements which contribute to the integrity of the whole aircraft are derived from these underlying assumptions". This is an acceptance of the probability argument without being specific about the precise values of probability that are acceptable.

The Society is represented on the following committees:- UK Flight Safety Committee, Civil Aircraft Control Advisory Committee, Air Registration Board, The Department of Trade and Industry (Operating Division) and Ministry of Defence (Procurement Executive) Joint Airworthiness Committee, AICMA Airworthiness Committee, Meetings of European Airworthiness Authorities, and Technical Co-operation Committee for All Weather Operations.

The opinion of the Director was that the work the Society's Committee on Airworthiness, and the Flight Operations Committee have some influence on government hazard control policy. This influence is probably exercised through representatives on these committees having their own contacts with similar interests in government departments. The Society does not employ a public relations specialist specifically to promote the Society's views in relation to the control of hazard.

In reply to the question about cases in which the Society has attempted to lead or influence government policy it was stated that "The Society has been represented in discussions on the following subjects: revisions and alterations of airways over the UK, Control Zones, Special Rules Areas, arrangements for high speed test flights, aircraft climb criteria, determination of UK transit levels and altitudes, visual flight rules criteria, aircraft accident investigation procedure, airmiss incidents, flight safety training, all weather operations, and frequent and regular participation in airworthiness discussions with the civil and military airworthiness authorities".

The Society's perception of the success that it had experienced by either having its advice accepted or seeing policy modified was expressed in the following way:- "Airworthiness requirements are constantly evolving as a result of the continuous contact which promotes identity of view between manufacturers and regulator authorities". The Society of British Aerospace Industries has the financial resources to support its activities, it has its own technical resources and can call on industry to help prepare the technical arguments it wishes to present. The Society has developed an extensive system of contacts with proximate policy makers, and can be considered to have an established place in the existing process of consultation.

The Air Registration Board

The Air Registration Board, was to a certain extent, an unusual
interest group, as it was financed by the industry and yet had
regulatory functions to perform. So at one level it was policy
making, and subject to the influence of the industry, and on
the other hand it tried to influence government policy on behalf
of the industry. The questionnaire was completed by Walter Tye
CBE, Chief of Executive of the Air Registration, and he
supported his answers with considerable documentary evidence.
The Air Registration Board consisted of eighteen members; four
representing constructors, four representing operators, four
independent members co-opted by the ARB, and one pilot, and one
independent person nominated by the Secretary of State. The
Board has 450 employees. Although the members were in the main
nominated by various sectors of aviation once appointed they
expressed their own personal viewpoint, but it was recognised
that they may well take into account the views of those who
have nominated them.

The policy of the Board was made collectively by the members of
the Board, and the staff prepared policy proposals for their
consideration. The staff ran several co-ordinating committees
with the aviation industry to explore reactions to proposed
technical policy and its effects. In general terms the
objective of the Board has been expressed as providing on behalf
of the public a measure of protection against aviation risks.

In reply to the question about what view the organisation had on
the way government policy should be developed on the control of
hazards the answer was given that "Where the control of hazards
is highly technical, and where practical implementation requires
responsible behaviour by the individuals working in industry,
there is a strong case for government to delegate as much as
possible of the responsibility to a body representative of
industry. The industry then exercises self-discipline which is
preferable to externally imposed enforcement".

The Air Registration Board's view on the acceptability of a
particular hazard being determined in probability terms was
expressed in the following way:- "When the probability of a
hazard can be numerically assessed, it should be so assessed.
In certain fields this is now technically possible. The
government and its agencies have a duty to seek to quantify the
consequences of decisions they take which affect hazards". This
is to an extent amplified by Tye, in reference 5, in which he
states that the acceptable probability of catastrophic failure
of an automatic landing system should be no greater than one in
10 million landings, and that for the systems analysis of
supersonic transports remote probability meant of the order of
10^{-5} to 10^{-7}. per hour.

The Air Registration Board's representation on government bodies
and committees related to the control of hazards was described
as being confined to aviation and hovercraft committees within
the Department of Trade and Industry and to International bodies

which determine international safety regulations.

The other methods the Board used to influence government policy on hazard control were described as resulting from the close working contact with the Division of the Department of Trade and Industry responsible for operational safety of aircraft. This working contact resulted in mutually acceptable interaction developing between the Division of the Department of Trade concerned with operational safety and the Air Registration Board The Board did not employ a public relations specialist to promote its views.

In reply to the question asking for details of cases where attempts had been made to influence policy the Air Registration Board claimed to have had an influence on the government decision to treat hovercraft as vehicles in their own right. The ARB urged that in technical detail the hovercraft should be treated on the basis of safety objectives rather than by arbitrary rules of thumb. In a contribution to a Royal Aeronautical Society Symposium, a copy of which was provided by the ARB (6), Mr. Tye explained that the ARB had made vigorous efforts to retain its autonomous status rather than becoming just a part of the Civil Aviation Authority, as described in earlier sections the ARB lost its independent status and is now part of the Civil Aviation Authority. *

The information available about the Air Registration Board suggests that the Board had adequate independence and the financial and technical resources to allow it to make a significant contribution to improving the safety of air transport, however the Board was not strong enough to preserve its original form of independent organisation in the face of the Government's wish to see it incorporated as part of the Civil Aviation Authority.

The British Air Line Pilots Association

The questionnaire sent to the British Air Line Pilots Association was completed by the General Secretary, who amplified his answer by providing an extract from the Association's Annual Report which gave details of the work of the Technical Committee. The Association has approximately 5000 members all of whom are private individuals. The members support the Association by subscriptions which varies according to salary rising from £12 per annum for those with salaries below £1,500 per annum to £48 per annum for those with salaries in the range £5,501 to £7,500. The members are considered to be responsible for policy either through voting directly at the Annual General Meeting or through their elected representatives at the meetings of the Central Board of Elected members or meetings of the Executive Council which has delegated powers from the Central Board. The Association's view of the way

* See page 93

government policy should develop on the control of hazards is
that they are not in favour of making the functions of flight
safety and airworthiness requirements the responsibility of the
same authority as has responsibility for the financial control
of such matters. This view appears very similar to the view
held by the Air Registration Board.

In reply to the question about the acceptability of determining
the importance of a particular hazard in terms of probability
the Association replied that in the absence of any better system
they considered the method acceptable. They added the rider
that a probability of $10^{-6} - 10^{-7}$ is the accepted yard stick in
the aviation industry.

The Association is represented on the following government
bodies and committees:- The Airworthiness Requirements Board,
Flight Safety Committee (UK), Flight Safety Foundation
(International), Airports Security Committee, The Flight Data
Recording Working Group, and the Airmiss Working Group.

Apart from representation on government bodies the Association
stated it would use pressure on government authorities when
discrepancies or deficiencies come to light. To this end it
would lobby MPs, and make its views known through the National
Press, the Association Journal, and Annual Technical Symposia.
To promote the Association's views it does employ a public
relations specialist.

The Association stated that it had attempted to influence
government policy in relation to the following:- flight time
limitations, all weather operations, aerodrome deficiencies
in navigational aids and facilities, helicopter legislation,
aircraft loading and aircraft security and hijacking. Rather
than specify cases where the Association's view was accepted
the Association replied that in most cases advice is not openly
or directly accepted, but in due course changes in policy are
frequently found to be in line to one degree or another with
suggestions or proposals pressed by the Association.

The Association although only representing a small but
important private individual section of the air transport
industry can stop virtually all British airlines flying by
calling its members out on strike and this is an important
resource. The members do make very large contributions to their
Association to give it adequate financial resources. The
Association does appear to fully exploit all techniques for
getting its views across, even to the extent of employing
Public Relations experts. One reason for the Association
being so widely involved in the government consultative
machinery may be its well developed Technical Committee. This
consists of seven representatives from BOAC, six from BEA, four
from the independent airlines, one representing professional
pilot training, and one representing the Civil Aviation Flying
Unit. The Committee meets monthly, operates a number of study
groups and has informal meetings with the Director General of
Safety and Operations, the Controller of the National Air

Traffic Control Services Division of the Board of Trade, the
Air Registration Board, and the British Airports Authority.
The study groups have covered: accident investigation,
airworthiness, Concorde, V/STOL, meteorology, navigation,
aircraft loading, legislation, airfield equipment deficiencies,
altimeter policy exemption of service pilots from Board of
Trade Examinations, flight recorders, hijacking and airport
security, and Luton airspace. Generally after examining the
problem they are concerned with they put forward papers which
contain suggestions for action to ameliorate the problem.
These papers if accepted can then go forward to become part of
the policy of the Association. The interesting feature of the
work of these study groups is that the pilots have to devote
their spare time to the furthering of this technical work,
from which to some extent other parts of the industry benefit.

The Guild of Air Traffic Control Officers

The Guild of Air Traffic Control Officers is concerned with the
professional conduct and efficiency of Air Traffic Control
Officers. There are 650 individual members plus 7 corporation
members. The Guild is administered by the Clerk of the Guild.
The activities of the Guild are financed by annual subscription
from the members. The policy of the Guild is developed by its
Council acting on the recommendation of the committees of the
Guild.

The Guild's view of the way government policy should be
developed on the control of hazards was expressed in the
following way: "The Guild keeps a constant watch on all
matters affecting the safety of air traffic and makes
recommendations to the Civil Aviation Authority or the
Government Department concerned when necessary". This suggests
that the Guild does not take the initiative in suggesting
entirely new policies that should be developed.

In reply to the question about evaluating the acceptability of
a particular hazard in terms of probability the Guild's view was
expressed as "not normally but this would depend on the hazards
and the odds".

The Guild is represented on the Air Safety Committee and the
Civil Aircraft Control Advisory Committee, but does not employ
public relations experts to bring the Guild's views forward.
The example the Guild gave of cases in which they had attempted
to influence government policy was related to the "Mediator"
air traffic control system. The Guild studied the proposals and
made criticisms and suggested improvements which were adopted.
Even the Guild's suggestions did not solve the problems with the
"Mediator" system as a report in the "Economist" (7) showed
that the Civil Aviation Authority admits the computer part of
the system will have to be thrown away, and possibly replaced
by a more efficient and cheaper American computer.

The Guild is a small organisation with small financial and
manpower resources. The Guild enjoys some interaction with the
Civil Aviation Authority and The Department of Trade and
Industry, sometimes this interaction can be inhibited because
the members are mainly public employees. Representation on
government bodies is limited, but attempts are being made to
extend the Guild's influence by holding symposia. In 1973 a
symposium on "Stress in Air Traffic Control" was held in which
the Medical Branch of the Civil Aviation Authority participated.

The Flight Safety Committee

Although the Flight Safety Committee did not complete the
questionnaire they did, in a series of letters, give a
considerable amount of information. The Committee was set up
to collect and disseminate flight safety information to foster
interest in and work for the design of flight safety in all
sections of aviation. It claims to be an independent body set
up by the Minister, and derives part of its financial support
from government sources. It does not see itself acting in the
role of an interest group, but merely working within the
framework of government policy to influence others towards an
improvement of safety levels. I feel this is strictly their
own wish not to define themselves as an interest group and is
a matter of their perception of their role.

The Committee meets 8 times a year for the purpose of exchanging
information and views on a variety of flight safety subjects.
The following organisations are represented on the Committee
and those marked with an * are the organisations which
supplement the government financial support of the Committee.

    *   The Civil Aviation Authority, Operations Division
    *   The Civil Aviation Authority, Airworthiness Division
    *   The Society of British Aerospace Companies
    *   British European Airways Group
    *   British Overseas Airways Corporation
    *   British Midland Airways
    *   British Insurance Association
    *   British Caledonian Airways
    *   British Island Airways
    *   Britannia Airways
    *   Court Line Aviation
    *   Dan-Air Services
    *   Donaldson International Airways
    *   Laker Airways
    *   Lloyd International Airways
    *   Lloyd's of London
    *   Monarch Airlines
    *   Tradewinds Airways
    *   Transmeridian Air Cargo
    *   Invicta Airways
        The Guild of Air Traffic Control Officers
        The General Aviation Safety Committee
        Commanders of Aircraft Representatives (BALPA and CAPAN
          nominees)

The Society of Licensed Aircraft Engineers and
    Technologists
The Ministry of Defence, Directorate of Flight
    Safety (RAF)

The main method by which the committee circulates the
information that it collects about accidents is by means of the
Flight Safety Focus Bulletin which contains accounts of
accidents and incidents from which lessons have been learned,
thus allowing others to learn from the mistakes. One thousand
copies per edition of this bulletin are published, and it is
published 8 times a year

There is always some reticence about exchanging information on
accidents and it is an important achievement of the Committee
that by respecting the confidential nature of these reports
it is still able to circulate details about incidents in a way
that people can learn from the analysis of the causes of
accidents.

The Committee says that it is used as a sounding board for
proposed legislation concerning safety, for example, at a
meeting in 1968 the Committee expressed concern at the number of
accidents that had occurred and were still occurring during
training flights, and submitted a paper to the Director of
Training and Licensing of the Department of Trade and Industry,
expressing views on the subject. The paper was to the
Assymetric Flight Training Working Group which the Director had
set up to examine the problem and the Committee was invited to
be represented on the Group. The Group made a series of
recommendations in its final report to the Director. Some of
these recommendations have already been implemented and others
are being worked upon. The Committee also organises a Flight
Safety Discussion Group which meets twice a year for discussion
on particular safety problems. Three expert speakers are
invited to address an audience from all sections of the industry
followed by questions and discussion from the floor. These
meetings are closed and representatives are encouraged to speak
freely without fear of being reported in the press and the
speakers can be as controversial as they wish. Typically these
meetings discuss problems such as pilot continuation training.
The Committee also liaise with Flight Safety Organisations in
other countries to help towards international exchange of
information.

Although the Flight Safety Committee does not see itself as an
interest group it is able to perform an influential role in the
development of air transport safety policy. Part of its success
is due clearly to the respect that it gives to the confidential
nature of the information that people present to it. Its
financial resources appear adequate, and the technical resources
at its command allow it to perform these functions quite
adequately.

Lloyds Aviation Underwriters Association

Lloyds Aviation Underwriters Association did not return the
questionnaire, because they did not see themselves actually
interested in influencing policy related to the control of such
hazards. They took the view that the safety record of a
particular risk is an important factor taken into account when
underwriters are determining the rate for the risk and to this
extent they exercise some influence on safety factors. The
Association was represented on the Air Registration Board and
individual members of the Association do take part in the public
discussions of safety matters at technical meetings such as
those held by the Royal Aeronautical Society. The role that the
underwriters perceive for themselves is thus of quite a low
order there being no formal attempt to establish an organisation
which could make a significant impact on safety policy, and
their representatives on such bodies as the ARB will, to a large
extent, be working independently on the basis of their own
particular knowledge of the problems discussed. However it
should be remembered that the rates they charge for insuring
aviation risks will have a significant economic influence on
acceptable hazard levels.

The Royal Aeronautical Society

The Secretary of the Royal Aeronautical Society declined to
complete the questionnaire as he did not perceive that the
Society had any influence on the formation of hazard control
policy. * The Society was founded 107 years ago, granted a
Charter of Incorporation in 1948 (8) and had a total membership
of 12,530 at the end of 1971. The prime objectives and purposes
of the Society as expressed in the Charter of Incorporation are:
"The general advancement of aeronautical art, science and
engineering and more particularly for promoting that species of
knowledge which distinguishes the profession of aeronautics".
The Charter also contains the statement that the Society should
"give the legislature and any departments thereof and public
bodies and engineering institutions and others, facilities for
conferring with and ascertaining the views of members of the
Society and other persons engaged in the profession of
aeronautics as regards matters directly or indirectly affecting
aeronautical art science and engineering and to confer, send
representatives to and communicate with all or any such
authorities and bodies in regard to the same". The Secretary
expressed the view that if the government were to ask the
Society for advice, then committees could be formed and then
advice given, and this had been done in the case of the Air Law
Group of the Society.

The Annual Report (9) of the Society showed that the Society is
financed mainly by subscriptions, and sales of publications.
The total income for 1971 was £212,436. Donations were also

*This ignores the fact that in the evidence the Society presented to the
 Edwards' Committee there was a section dealing specifically with safety.
 See page 44

received from the British Aircraft Corporation Limited, British
Aviation Insurance Co.Limited, Hawker Siddeley Aviation Limited,
Ministry of Aviation Supply, the Society of British Aerospace
Companies Limited and Westland Aircraft Limited. The income
account showed that for the year up to the end of 1971 donations
to general funds amounted to £3,197. Salaries, wages, National
Insurance and Pension Premiums for the Society for the year to
the end of 1971 amounted to £93,770.

The Society sets academic standards that applicants for
corporate membership have to satisfy. The importance the
Society attaches to the training and education of engineers in
general and to the aircraft industry in particular is underlined
by the fact that the Society is a member of the Council of
Engineering Institutions, which is the organisation nationally
responsible for administering the system of qualification for
Chartered Engineers.

The Council of the Society, which is the policy controlling
organisation of the Society, is made up of people in senior
positions, that can probably be classed as proximate policy
makers.

Within the Society it is divided into seven specialist groups
which are: Agricultural Aviation Group, Air Law Group, Air
Transport Group, Historical Group, Man-powered aircraft Group,
Management Studies Group and the Test Pilot's Group.

During 1971, the Air Law Group held a Symposia on Manufacturer's
and Equipment Suppliers' Liability for Aircraft, Industrial
Relations in the Aviation Industry, and the Hague Convention on
Hi-jacking of Aircraft. A discussion meeting was held on the
Legal Aspects of International Co-operation on Aircraft Design
and Production. Also a joint meeting with the American Bar
Association was held to discuss aircraft noise and the
selection of airport sites.

The Society is represented on British Standards Institution's
Committees and on the International Standardisation Organisation.
It is also represented on the Courts of several universities and
examining bodies. The Annual Report of the Society (9) gives
no indication that the Society is represented on any government
committees or bodies.

A unique operation that the Royal Aeronautical Society carries
out is to operate the Engineering Sciences Data Unit Company
which provides data on approved methods of calculation, and
physical properties. These data reports cover: aerodynamics,
fatige, fluid mechanics, stress analysis and physical data.

FACTORIES

At the time this study was made many organisations had been
active in presenting their views on the way government policy
should develop on the control of hazards to Lord Robens'
Committee on Health and Safety at Work. Seven of the

questionnaires returned reported that evidence had been
submitted to Robens' Committee. So views on factory hazards
obtained by the questionnaire cover a somewhat wider range of
organisation than originally expected.

Questionnaires that were anticipated would be particularly
concerned with factory hazards were those sent to the
Confederation of British Industry, the Institution of Mechanical
Engineers, and the Royal Society for the Prevention of Accidents
and the Amalgamated Union of Engineering Workers. These four
organisations gave complete answers to the questionnaires.
Additional information on evidence given to the Robens'
committee was obtained from the Association of Municipal
Corporations, The National Union of General and Municipal
Workers, the Institution of Professional Civil Servants and the
British Medical Association.

The Confederation of British Industry

The CBI did not return the questionnaire, but gave me an
interview with two assistant Directors who answered all the
questions and provided me with considerable information about
the organisation of the Confederation and the way its policy is
formed.

The Membership of the Confederation on 31 December 1970
consisted of 11,436 Industrial Companies, 193 Commercial
Companies, 15 Public Sector members, and 217 Employers'
Organisations and Commercial Associations. The staff of the
Confederation number about 380. The total income of the
Confederation for 1970 was £1,234,832 of which members
subscriptions amounted to £1,122,441 the rest being made up
mainly of income from rents and investments.

The Confederation is controlled by a Council elected partly by
a system of regional selection, and partly by co-options by the
Council itself. The Council holds eleven regular meetings a
year. The Council is supported by thirty standing committees,
which are detailed in Appendix VI. An example of how CBI
policy is formed was given with reference to the way the CBI
case for presentation to the Robens' Committee was prepared.
The CBI staff after consultation with employers organisations
and other representative organisations, prepared papers which
were reviewed and developed in the light of comments by a
special working party that had been established for that
specific purpose by the Industrial Relations Committee. The
final versions of the papers were first approved by the
Industrial Relations Committee, and then passed to CBI Council
for approval before being submitted to the Robens' Committee.
The Confederation generates policy in response to several kinds
of stimuli, sometimes in response to approaches through members,
sometimes as a result of approaches by the civil service, and
on other occasions as the result of research and evaluations of
situations by CBI staff. This flexible approach of the CBI to
the development of policy means that policy develops in an
incremental way.

The views that were expressed on the way government policy should develop on the control of hazards made the following five points:-

1. There is a need to prevent overlapping jurisdiction of government Departments which may result in confusion and even in possible conflict resulting in situations which may themselves be dangerous.

2. That industry should be encouraged to continue to develop voluntary systems of joint safety consultation.

3. The Factory Inspectorate's role should become more that of advisers on safety rather than enforcers.

4. New legislation should be introduced, which amplifies existing legislation and sets out employer's and employee's duties in simple straightforward terms.

5. Proof of conformity with safety standards by self-certification is preferred although it is accepted that third-party certification may be necessary in some cases. It was further considered that existing Testing Houses and Approval Boards are adequate for the third-party certification required and that no new organisation should be necessary.

These views could be taken to be the goals of the CBI in relation to hazards control in factories.

The view expressed on the acceptability of using probability techniques to evaluate particular hazards was that, the CBI appreciated it could be a useful approach in some cases and they could foresee problems in its universal application.

The extensive range of government bodies on which the CBI is represented is shown in Appendix VII. Apart from the day to day contact with government departments that the CBI has, it holds discussion and lecture meetings for interested parties and press conferences to put its view. The CBI has an Information Directorate to help disseminate information about the Confederation's views and how government proposals would affect them.

In the recent past the CBI has attempted to influence government policy in relation to roads, air transport, industrial safety, and clean air. The CBI was concerned that the road programme was not adequate in the sense that the access to ports and airports was not adequate for the needs of industry. But improving roads does have safety overtones. The CBI did present its views to the Edwards' Committee when the White Paper on the future of civil aviation was being prepared. Their concern was that not sufficient allowance was being made for the protection and representation of the consumer. The White Paper finally published, was considered acceptable.

The CBI was consulted over the drafting of exemption orders under the 1969 Clean Air Act and the preparation of the explanatory memoranda on the Act that the Ministry published.

In the industrial safety field the CBI made representations against the introduction of compulsory joint consultation on safety. It is possible that the representations made were influential in the Minister deciding to set up the Robens' Committee which has been mentioned above, and the impact of this evidence on the Robens' Committee is discussed in the body of this study.

The Confederation has extensive technical and financial resources but cannot control the production of its members in the way a union can. It has well developed formal and informal contacts with the government. It represents the opinion of most of the industrial companies in the country. However, if some sector of industry had its own specialist organisation it would not interfere. The aircraft and motor industries were cited as two industries with their own specialist organisations. By the nature of its method of operation the approach to policy development is essentially incremental, although, from the evidence the CBI has presented, it is possible to detect the synoptic view that as far as possible industry should be allowed to regulate its own affairs on a voluntary basis.

The Institution of Mechanical Engineers

The organisation of the Institution of Mechanical Engineers was described in the section on road transport, and it was seen that the only influence they bring to bear on government policy is indirectly through the meetings and discussions that they hold. * Two discussions were held during the 1971-72 session that have a general bearing on hazard control in factories, these discussions were entitled:- "Noise and Vibration Engineering" and the "Environmental future and its dependence on engineering education". The influence of the Mechanicals in this field, is similar to the indirect influence which it has in road transport, and results from providing a forum for discussions on technical aspects of the problem at which solutions and views can be fairly openly explored.

The Royal Society for the Prevention of Accidents

The Royal Society for the Prevention of Accidents was also described in the Road Transport Section. In 1970-71, Lord Robens was a Vice-President of the Society, so clearly he would be aware of the views of the Society and be able to take them into account in the deliberations of his committee on Health and Safety at Work. The annual report stated that evidence had been presented to the Robens' Committee, but no details of the nature of the evidence was given. The Annual Report stressed that training in safety is a very important activity in the Industrial Safety Division of the Society. The educational function is performed by the Society running courses and supplying technical advice and information.

* See note Page 90

The comments made on the Society in relation to road safety
appear to be applicable to the role that the Society plays in
relation to factory safety, that is it has well developed
contacts but technical resources are somewhat limited.

The Amalgamated Union of Engineering Workers

This union has a membership of 1,240,000 and a staff of 350.
Members pay subscriptions of between 5p and 25p per week so the
union income is probably in the region of £3-14 million per
year.

The policy of the Union is formed by proposals first approved by
Branch Meetings and then passed to District Committees and then
finally subject to approval by the Annual Rank and File Delegate
Conference.

The Union's view on the way government policy should be
developed on the control of hazards are: there should be
stricter control by government agencies and enforcement of
existing legislation, there should be an occupational heatlh
service, and a co-ordinated environmental and pollution control
to monitor dangerous substances. The Union also accepted the
use of probability techniques to assess the acceptability of
particular hazards.

The Union is formally represented on a number of official joint
standing bodies. At the informal level the union is prepared to
put pressure on specific industries by using industrial action.
It considered it had attempted to influence policy in relation
to: joint consultation, hazards of mineral oils, dangers from
power presses and abrasive wheel regulations, but the Union felt
that it could not be specific about the extent to which it had
influenced policy.

One practical way in which the Union tries to reduce hazards is
by keeping its members informed about the nature of hazards and
the way they can be reduced. It does this by circulating a
quarterly Industrial Health add Safety Bulletin to its members.
Recent issues have dealt with subjects such as deafness at work,
vibration syndrome, hazards of working with acetylene, skin
cancer, treatment of phenol splashes and mercury hazards. This
bulletin is prepared by the specialist Industrial Health and
Safety Department of the Union.

The Association of Municipal Corporations

The Association of Municipal Corporations, which is described at
length in the air contamination section, also presented evidence
to the Robens' Committee. They suggested that a single statute
should replace the Factories Act 1971 and the Offices Shops and
Railway Premises Act 1963. The Association also submitted that
the present allocation of responsibilities between the Factory
Inspectorate and local authorities is illogical, and suggested
that local authorities should be responsible for dealing with

the general health and welfare provisions of the Factories Act,
whether or not mechanical power is used, and also for the
administration of the 1963 Act in all non-industrial premises,
including local authority offices and schools and Crown Premises.
This proposal was intended to eliminate the overlap between the
duties of Public Health Inspectors and Factory Inspectors.

The National Union of General and Municipal Workers

The reply that the National Union of General and Municipal
Workers gave to the questionnaire was mainly related to nuclear
power reactors and their reply is discussed in that section,
in addition they commented on factory safety. The view was
expressed that Factory Inspectors should, where possible, have
experience of the industries they cover and of their specific
safety and health hazards. New processes should be more
carefully researched and inspected by law to establish their
long and short term health hazard to the human race. The Union
had made representation to the Robens' Committee but gave no
details of the evidence presented.

The Institution of Professional Civil Servants

The Institution of Professional Civil Servants, the union whose
membership includes members of the Factory Inspectorate also
submitted evidence to Lord Robens. The Institution's overall
organisation together with their reply to the questionnaire is
described in the section dealing with nuclear power reactors.
After the Robens' Committee was set up in August 1970, the
Institution set up an ad hoc working party consisting of
representatives of all branches of the Institution where there
were appropriate inspectors in the membership. In December
1970, the Institution submitted a paper to the committee which
made the following recommendations:-

a)   legislation should be co-ordinated and unified as
     far as possible.

b)   there should be a co-ordinated enforcement and
     advisory service, with possibly all the main
     inspectorates attached to one Ministry.

c)   properly staffed inspectorates with efficient
     support teams were essential to maintain a
     proper service.

d)   fines should be raised to a realistic level and
     court orders used more widely.

e)   an appropriate diploma in occupational health
     and safety should be introduced for future
     Factory Inspectors and for suitable candidates
     from industry.

f)   the inspectorates should have at their disposal
     co-ordinated research facilities.

g)   Inspectors should be empowered to apply for court
     orders where the public safety is endangered by

radiation and toxic waste.

Subsequent to the submission of this evidence the Institution
was asked to give oral evidence with particular reference to
legal enforcement.

The British Medical Association

The British Medical Association, which is described in the
section dealing with air contamination, also presented evidence
to Lord Robens' Committee. In their evidence they drew
attention to the fact that of 2000 factories with more than 500
employees, only some 1300 had the services of a part time or
full time doctor. The Association considered that there was
little need for argument about the importance of expanding the
occupational health service, as for every 1 day lost in strikes,
about 10 days are lost by industrial injury and disease and
approximately 100 days are lost by ordinary illness. The
Factory Inspectorate was concerned to prevent occupational
accidents and disease. The National Health Service treats
illness, and within the service the Public Health branch is
concerned to prevent ordinary illness. Outside the privately
provided occupational health service, there was little or no
provision for controlling and minimising the effects of
ordinary illness on mens' ability to work, for assessing the
effect of illness and physical handicap on capability for work,
for medically supervising the return to work of the recovering
patient, for minimising the bad effect of work on the partially
fit and, even more importantly, for maximising its good effect.
To satisfy this need, the Association suggested two ways in
which an occupational health service might be established. First,
the provision by government, at government expense, of a service
which would be organised outside industry and imposed on it.
Second, by making it a legal requirement that industry make
provision for its own health service. Coupled with this need
for an occupational health service, the Association saw the
need for an Occupational Hygiene Laboratory Service possibly as
a development of the work of the Public Analysts.

The Association also expressed the view that if Britain joins
the European Economic Community, the country will be found to be
deficient in occupational health services compared with other
countries in the Community.

NUCLEAR POWER REACTORS

Nuclear power reactors are the smallest and most tightly
controlled of the five activities considered. This means that
there are few alternatives to the interest groups selected for
study. In this sense, it was unfortunate that only two of the
five organisations in the nuclear power field answered the
questionnaire fully, and that these two organisations were both
trade unions. To prevent this part of the study being biased
towards trade unions, published information has been used to
show something of the views of the other types of organisations.

Questionnaires were sent to the Central Electricity Generating
Board, the two nuclear power plant contractors TNPG and BNDC,
the Institution of Professional Civil Servants and the National
Union of General and Municipal Workers. The Institution of
Professional Civil Servants and the National Union of General
and Municipal Workers completed the questionnaire fully. The
CEGB, the main owner of nuclear power stations in this country,
in their letter declining to complete the questionnaire,
expressed the view that most of the required information was
published in their Annual Report.

The Annual Report for 1970-71 (10) shows that of the Board
employees 5,068 were classified as radiation workers during 1970,
and expressed the view that the nuclear health and safety record
of the Board's nuclear power stations and laboratories continued
to be satisfactory. It was also mentioned that the results of
environmental surveys of radioactivity carried out around the
stations were reported to the Local Liaison Committees by
representatives of the authorising Ministries, and that the
results of these surveys were entirely satisfactory. It was
reported that plans for dealing with emergencies had been
closely re-assessed and that rehearsals continue to be carried
out to the satisfaction of the Nuclear Installations
Inspectorate of the Department of Trade and Industry.

To illustrate how opinions and policy on the control of nuclear
power reactor hazards have developed and the various interest
groups have interacted, the siting of reactors is considered,
in the same sense that it was considered in the section on
policy formation. At the IAEA Symposium (11) on siting in 1967,
the Central Electricity Generating Board, the nuclear power
plant contractors, the United Kingdom Atomic Energy Authority,
and the Nuclear Installations Inspectorate presented papers on
the siting of reactors. The CEGB presented a paper in which the
risk to individuals living near a nuclear power station was
discussed in probability terms. In a paper by Charlesworth and
Gronow of the Nuclear Installations Inspectorate, the experience
of the Inspectorate in evaluating reactor sites was reviewed.
The method of evaluating sites they described was based on
comparing site risk factors, which were calculated by summating
the product of the population in various radial sectors around
the site and weighting factors. The weighting factors being
selected so that they represented the change of risk with
distance from inhalation of iodine 131. Charlesworth and Gronow
drew attention to the fact that siting alone can never be relied
upon to safeguard the public in the UK, and that great
attention must therefore be concentrated on examining the means
whereby a large release is precluded, not only at the inception
of the project and during construction but also in the course of
operation of the plant. Some doubts were expressed about
advisability of trying to quantify risk and to express it in
terms of probability.

At this symposium, Mr. Farmer of the United Kingdom Atomic
Energy Authority presented a paper showing that it was a

practical proposition to determine for a nuclear reactor the
characteristic spectrum of failures probabilities associated
with it. From this failure/probability characteristic of the
reactor it was postulated that it is possible to decide if the
reactor is acceptable or whether it needs additional safety
features built into it to make it acceptable. The wider
implications of this technique are discussed in Appendix II.

Two papers were presented by nuclear power plant contractors at
the symposium, one by L Cave and R E Holmes of the Atomic Power
Constructors Limited, on the suitability of the advanced gas-
cooled reactor for urban siting, and the other by
J M Yellowlees and T W Spruce of the Nuclear Power Group Limited
dealing with the safety features of the Hinkley Point'B'AGR
pressure vessel and penetrations. The paper by Cave and Holmes
did show how, using the Farmer approach, a particular type of
reactor could be shown to be suitable under certain conditions
for urban siting. The paper by Yellowlees and Spruce dealt with
the technical design aspects of the safety features of the
Hinkley "B" AGR pressure vessel, and did not get involved with
either the question of siting or probability of failure.

Two years later, in 1969, the British Nuclear Energy Society
held a symposium on safety and siting. The organising
committee of this symposium consisted of representatives from
the Central Electricity Generating Board, the nuclear power plant
contractors, the United Kingdom Atomic Energy Authority, and the
Nuclear Installations Inspectorate. As already mentioned in
the section on policy formation *,at the opening of this
Symposium,Sir Owen Saunders, Chairman of the Nuclear Safety
Advisory Committee drew particular attention to the papers
dealing with the application of probability analysis. This
suggests that the Nuclear Safety Advisory Committee accepted the
use of probability analysis, which had first been suggested
could be applied to the nuclear industry two years before. At
the symposium the papers by the Nuclear Installations
Inspectorate, CEGB, UKAEA, and the power plant constructors
suggest a considerable degree of agreement between these
organisations on safety policy.

The Institution of Professional Civil Servants

The Institution of Professional Civil Servants, the trade union
that represents professional civil servants had a total
membership of 94,940 at the end of 1971. This membership covers
most factory, alkali and nuclear installations inspectors, and
represents the professional staff of the United Kingdom Atomic
Energy Authority. Members pay an annual subscription of £8.40.
The policy of the Institution is formed by the Annual Conference
of Branch Delegates and by the National Executive Committee
working within guide lines laid down by the Annual Conference.
Although the Institution is not represented formally on any
official government bodies or committees concerned with the
control of hazards, it has made representations on behalf of its

*See page 54

members and it has presented evidence to the Robens' Committee.
In reply to the questionnaire, the view was expressed that the
Institution found itself in a difficult position to influence
government policy on hazard control as its members are working
in a professional capacity to control hazards. The evidence
that the Institution presented to the Robens' Committee (12)
is interesting in that the proposals it makes for controlling
hazards at work are modelled on those existing in the nuclear
industry. This suggests that the Institution is satisfied with
the arrangements that exist in the nuclear industry and that
they would be an improvement on the arrangements that exist in
other fields.

The answers that the IPCS gave to the questionnaire give the
impression that although the IPCS has considerable financial
resources and can call on leading experts among its members,
its actions are to some extent inhibited because its members
are government employees. The main impact that the Institution
has is through negotiating machinery that it is represented on,
and which plays an important part in deciding how policy can
and will be implemented with regard to staffing and organisation.
This machinery does, however, give the Institution access up to
Ministerial level. Judicious use of this machinery at times
when new legislation is being considered, which may affect IPCS
members, can give the Institution some impact on policy.

The National Union of General and Municipal Workers

The answers given by the National Union of General and Municipal
Workers were quite comprehensive. The Union has a membership
of about 850,000 members, who pay subscriptions of either £9.36
or £6.24 per year, and a total staff of 500, 150 of whom are
full time. The policy of the Union is formed at the Annual
Congress, attended by elected delegates from each region. The
views of the Union on the way government policy should develop
with respect to the control of hazards has already been related
with respect to factories, namely that inspectors should be
experienced in the industry and that new processes should be
carefully researched before they are approved. Similar arguments
could be applied to any industry, and this view appears to
endorse the situation that exists in the nuclear industry.

The view expressed for the Union was that no risk was ever
small emough for it to be accepted and that constant vigilance
and improvement will always be required so long as there is any
chance of accident.

It was claimed that the Union makes some impact on the control
of hazards through its representations on joint safety committees
in the nationalised industries. Government policy on the control
of hazards is influenced by representations that the Union makes
through the TUC and other centralised bodies such as the
Confederation of Shipbuilding and Engineering Unions. The Union
does not employ public relations specialists to promote its
views.

One case in which the Union claimed to have led or influenced government policy was when the Union members in the atomic energy industry were exposed to plutonium over-burden. Negotiations took place with the UKAEA on the need for more stringent control as well as compensation. It was claimed that these negotiations helped to ensure a very high standard of safety in the UKAEA.

The Union has, like other unions, extensive financial and technical resources at its command. It does not consider itself restricted in the action it can take in the same way that a union representing only civil servants does. On the other hand, the contact with proximate policy makers does not appear to be as highly developed as say that of the IPCS and reliance is placed on central representation through bodies like the TUC and the Confederation of Shipbuilding and Engineering Unions. Representation through the TUC is not at present open to the IPCS as they are not affiliated to it.

AIR CONTAMINATION

Of the five activities considered, air contamination is the one that makes a universal impact on the health of the nation. It is also an activity with very widely distributed responsibility for its control.

The response to the questionnaire was satisfactory in this area. Seven of the ten questionnaires sent out were answered fully. The organisations that responded were: The Confederation of British Industry; The British Medical Association; The Institution of Civil Engineers; The Association of Public Health Inspectors; The National Society for Clean Air; The Civic Trust and the Association of Municipal Corporations. The Clean Air Council declined to complete the questionnaire. The Institution of Public Health Inspectors and the Urban District Councils Association did not reply at all. In view of the wide distribution of the interests of the organisations that did respond, no further comment is considered to be necessary on the role played by the last two organisations. The reasons for the Clean Air Council not completing the questionnaire are worth noting.

The Secretary of the Clean Air Council very politely declined to complete the questionnaire as the Council's proceedings are confidential. The Clean Air Council was appointed to advise the Secretary of State for the Environment on air pollution matters, a Minister takes the chair at its meetings and the Secretariat of the Council is provided by the Department of the Environment. The Council's reply shows how the nearer an organisation is to the proximate policy maker the harder it often is to determine the nature of the advice they are giving policy makers.

The Confederation of British Industry

The Confederation of British Industry, whose organisation is described in the section dealing with factory hazards, summarised its views on air pollution in the following way in the Annual Report for 1970:

The British practice of subjecting discharges of liquid or gaseous effluents to individual consents is a procedure that is both practical and flexible, and much preferable to the rigid approach commonly adopted abroad.

Conditions imposed on discharges of industrial effluent must not only be reasonable but technically and financially capable of achievement. Conditions must be related to the circumstances of each discharge and of the receiving source.

Modernisation of industrial operations and increased investment in new processes and plant, will lead to a significant reduction in pollution from industrial sources.

The international aspects of abatement of pollution are becoming increasingly important. The CBI welcomes international collaboration on problems associated with environmental pollution but is satisfied that arbitrary and rigid international standards are completely unworkable.

During the interview with the CBI officials they expressed the view that:- a pollution tax would be an unsatisfactory solution to the problem, as it could lead to odd results in the distribution of industry. They also considered that pollution control in Britain should not be out of line with what is done in other countries, otherwise British industry could be adversely affected.

To help spread understanding of pollution problems the CBI held a one-day conference on technology and the environment in 1970. This conference was considered to be part of the CBI's contribution to European Conservation Year.

During 1970, the CBI was consulted over the drafting of two new orders under the Clean Air Act 1968. These orders were concerned with laying down limits for emission of grit and dust from furnaces, and giving local authorities powers to call for measurement of grit and dust emissions from the extended range of furnaces brought in under the Act. The CBI was also consulted over the drafting of a new Alkali Works Order which introduced amendments to scheduled works and to the list of noxous and offensive gases.

The British Medical Association

The British Medical Association gave their answers to the questionnaire at an interview, so it was possible to obtain some supplementary information. The prime object for which the

Association was established in 1832 is stated as being:-
"To promote the medical and allied sciences, and to maintain
the honour and interests of the medical profession". Doctors
do not have to belong to the Association and the Association is
not a qualifying body.

The Association has a membership of about 70,000 and the
subscriptions from these members yielded just over £655,000 in
1971. The staff of the Association, apart from those concerned
with the Association's publications, is about 300. Twenty-
three of the staff are in the administrative grades, which are
functionally equivalent to the administrative grades in the
civil service.

The Council of the Association, which is 50 strong, is elected
so as to ensure that the geographical spread of the various
divisions is properly represented. It is the Council of the
Association, acting on the advice of the various standing
committees, which is the central executive and policy forming
body of the Association. It is the sub-committee or working
parties, established by the standing committees, that generally
prepare the advice that is passed to government bodies,
considering changes to or the introduction of new legislation
that may affect medical practices.

The view expressed on the way government policy should be
developed on the control of hazards was stated quite simply to
be that "if it is shown that a significant risk exists
appropriate action should be taken".

From this followed the view that wherever possible a hazard
should be eliminated, but where it is not possible to eliminate
the hazard there is no alternative to determining the
acceptability of the hazard in terms of probability.

The Council of the Association is from time to time asked by
Secretaries of State to nominate representatives to various
government bodies. Contact with the government is also
developed through MPs, and maintained through close association
with the civil service. An interesting feature of the
Association is that it maintains a fund that members of the
medical profession can call on for support if they are seeking
election to parliament. The fund is operated without any party
bias.

The Association does have a public relations department and this
is seen as helping the Association in two ways. First, it helps
to make public the Association's views and secondly, it keeps
the Association aware of medical matters that are causing public
concern. This second function of the public relations
department is important in that it allows the association to
develop opinions at an early stage in the development of a
problem.

With regard to air contamination, the Association recommended to
the government in 1963 that it should speed the implementation

of the Clean Air Act and improve the supply, efficiency and cost of smokeless fuels and give local authorities more financial assistance to increase their staff for this purpose. The Association also presented evidence to Sir Eric Ashby's Working Party preparing the United Kingdom submissions for the United Nations Conference on the Human Environment in Stockholm in 1972. The evidence dealt with various forms of pollution including air pollution and presented views on the long term dangers and suggested priorities for inquiry and action, and the scope for international agreement. In the evidence to the Robens' Committee, the Association recommended that the legislation to protect the public against increasing pollution from noise and exhaust fumes on both aircraft and road vehicles should be strengthened.

The British Medical Association is a well established interest group with a long tradition of formal and informal interactions with the proximate policy makers. It is a major source of medical advice and has considerable financial resources. The approach to the development of policy is incremental in that problems are tackled as they arise.

The Institution of Civil Engineers

The Institution of Civil Engineers, which was described earlier in the section dealing with road transport, is also concerned with the environment. The Institution set up an ad hoc committee chaired by Sir Kirby Laing to prepare and present its views on the human habitat to Lady Dartmouth's Working Party which prepared a report for the UN Conference on the Environment in Stockholm in 1972. This demonstrates the Institution's willingness to use its resources to influence policy on environmental matters, and that its interaction with the proximate policy is sufficiently close for it to be able to respond quickly enough to provide the evidence required in time to have an impact on the report of Lady Dartmouth's Working Party.

The Association of Public Health Inspectors

A smaller and more specialist professional body with special interest in air contamination is the 87 year old Association of Public Health Inspectors. The Association has a membership made up of private individuals of over 6,000. In the year ended 31 March 1971 the subscription income amounted to £25,621. The Association has a staff of eleven. In the year 1970-71 Mr. Eldon Griffiths, Conservative Member of Parliament for Bury St Edmunds, and Parliamentary Secretary to the Department of the Environment was President of the Association. Among the 21 Vice-Presidents of the Association were 7 members of Parliament, one of which was Mr. Peter Walker who was then Secretary of State for the Environment. The basic goal of the Association is promotion of the acquisition of knowledge special to the role of Public Health Inspectors.

The policy of the Association is made by an elected General
Council, in consultation with the various centres and branches
of the Association.

The view expressed on behalf of the Association on the way
government policy on the control of hazards should be developed
was that the Association was broadly satisfied with government
policy and the progress being made to reduce air pollution, but
that more comprehensive and effective legislation on the control
of noise was required.

In answering the question of whether or not the acceptability of
a particular hazard can be determined in terms of probability,
it was stated that the evaluation of a hazard should take into
account effects on health suspected or proved, practicability of
remedial measures, and cost of remedial measures.

The Association, as such, is not represented on any government
bodies related to the control of hazards. But public health
inspectors are in a personal capacity members of the Clean Air
Council and the Noise Advisory Council. However, the
Association does make its views known by: representations to and
discussion with government departments, presenting evidence to
official committees and working parties, and it holds
conferences on specific and general environmental health topics.
To help promote the Association's view it does employ a public
relations officer.

The Association has a Legal and Parliamentary Committee and
during the year 1970-71 this Committee took the following action
related to Clean Air:-

1)   Discussions with government departments on the
      availability of solid smokeless fuels.

2)   Observations were sent to the Ministry of
      Housing and Local Government on the proposals
      for the regulation of emission and measurement
      of grit and dust from furnaces. The Association
      also attended an informal meeting at the Ministry
      of Housing and Local Government to consider
      future policies on air pollution measurement.

3)   The draft order under the Alkali Works Regulation
      Act 1906 adding to the list of scheduled works
      petroleum, mineral and certain other processes
      was strongly opposed. The Association reiterated
      its policy that local control should be retained
      and that the proposal could reduce efficiency
      and the service to the community.

Although small, the Association can call on resources with
specialist technical knowledge of the problems of air
contamination. The Association has informal contacts with
government departments, and through its Vice Presidents it has
considerable lines of contact to several Members of Parliament
and Ministers.

The National Society for Clean Air

The National Society for Clean Air, which has developed from
the Coal Smoke Abatement Society founded in London in 1899, has
a wide range of membership. The membership of 1,200 is made up
of private individuals, industry, local authorities and learned
societies. The subscription and donation income amounted to
£19,000 to 31 March in the year 1970. The Society has a staff
of 7 at its headquarters. The objectives of the Society are to
promote an informed public opinion on the value and importance
of clean air and to initiate, promote and encourage the
investigation and research into all forms of atmospheric
pollution in order to achieve its reduction or prevention.

The policy of the Society is made by the Council elected from
the members, and including nominated representatives from
"National organisations" such as the Gas Council. In forming
policy the Council is assisted by the various committees of the
Society.

On the question of how the Society would like to see government
policy develop on the control of hazards the following answer
was given:- "The Society would like to see a general tightening
of controls on air pollution. The smoke control programme is
not going as fast as it should. Stricter limits are required
for some emissions from industry and there is an urgent need
for legislation to control emissions from motor cars".

The Society did not comment on the use of probability techniques
for assessing the acceptability of hazards, as it was considered
that this technique had not found significant application in
the air pollution field.

The formal interaction that the Society has with proximate
policy makers is through its representation on:- the Clean Air
Council, the Standing Conference of Co-operating Bodies,
Department of the Environment Working Parties on Odours and
Grit and Dust, and on Sir Eric Ashby's Working Party on
Pollution. At the informal level the Society enjoys good
relations with government departments concerned with clean air,
and has access to the Minister if required. The Society does
not employ a public relations expert, its views are put forward
by means of lectures, conferences and publications.

The cases the Society cited as being those in which they had
been able to influence Government policy were the Clean Air Act
of 1956 and 1968. The 1956 Act was originally sponsored by the
Society and introduced by Gerald Nabarro as a private member's
Bill. It was then taken over by the Government and duly became
law. The Society was consulted by the Government about the 1968
Act and later about some of the specific regulations issued as
a result of that Act.

The Society is a single interest group, its resources are that
its membership includes a large proportion of the local
authorities that have to administer a significant part of the

legislation concerned with the control of air contamination, and that it developed formal and informal links with the proximate policy makers.

The Civic Trust

The Civic Trust is rather different in character from the other interest groups described, in that it was founded by a Member of Parliament with extensive ministerial experience, namely Mr. Duncan Sandys. The Civic Trust is recognised as a charity, and its income is derived mainly from covenants contributed by industry and commerce.

The objectives of the Trust are to stimulate the improvement of the quality and appearance of towns and cities and to protect the countryside.

The policy and activities of the Trust are directed by a Board of Trustees which is presided over by Mr. Duncan Sandys. Nine of the twenty members of the Board of Trustees are Members of the House of Lords. The policy is built up from discussions involving the staff and the President, and often in the light of representations and views submitted by local authorities other public bodies, voluntary organisations and members of the public.

The Trust had no general views on the way government policy on hazard control should develop, nor had the Trust any views on the use of probability techniques as a method of determining the acceptability of hazards.

The only government body related to hazard control that the Trust was represented on was the Noise Advisory Council. The Trust uses conferences, correspondence and discussions with Ministers of the Crown and Government officials to influence Government policy. With a president who is so well known in parliamentary circles, the Trust should have easy access to Ministers. The Trust does not employ a public relations specialist to promote its views.

In the memorandum on heavy lorries that the Civic Trust submitted to the Minister of Transport in October 1970, the Trust recommended that there should be stricter enforcement of the regulations governing vehicle exhaust fumes.

From the above, it appears that the Civic Trust is not a major interest group concerned with developing policy for controlling air pollution, but because of the high level of contact that it enjoys in the Houses of Parliament the views it does express probably have significant impact.

The Association of Municipal Corporations

The Association of Municipal Corporations gave their answers to
the questionnaire at an interview in which they amplified
particular aspects of the Association's organisations and views.

The object of the Association which was founded in 1880 is to
watch over and protect the interests, rights and privileges of
municipal corporations, as they may be affected by Public Bill
legislation, or by Private Bill legislation of general
application to boroughs; and in other respects to take action in
relation to any other subjects in which municipal corporations
may be interested.

The Association is supported by subscriptions which are
proportioned to the size of the Corporations. All Municipal
corporations do belong to the Association, and in 1971 total
subscriptions amounted to £178,000.

The total staff of the Association is 36, of which seven are
counted as senior officials. Five of these senior officials
have had legal training, one was a finance specialist and the
seventh was a generalist. The work of the Association's
committees is mainly undertaken by staff of and paid for by the
Corporations.

Policy of the Association is made by the Council acting on the
advice it receives from its committees. The council is a large
body consisting of about three representatives from each
borough, and meets four times per year. There are thirteen
committees of the Association, of which the Health Committee is
the committee relevant to the present discussion.

Two aspects of air contamination were dealt with by the Health
Committee during 1970/71. First the Association expressed
strong reservations about the scheduling of mineral processes
under the Alkali,and Works Regulations Act 1906 because they
considered they could readily be supervised by local
authorities who are in general in a better position to do so
than the Alkali Inspectorate. Second, the Committee urged that
the cost limits for grants towards smokeless appliances laid
down in 1966 should be increased. During the interview the
Association representative expressed the view that the
improvement in air quality since the Clean Air Act of 1956 was
one of the successes of municipal corporations administration.
The areas that had not fully implemented the Act were mainly
Urban District Councils and Councils responsible for country
areas.

The Association has over the years developed a close working
contact with government departments and with members of both
Houses of Parliament. Contact with members of Parliament was
facilitated by the fact that the Association has 10 Vice
Presidents who are members of the House of Lords and 20 Vice
Presidents who are members of the House of Commons.

The resources of the Association are extensive in the sense
that it can call upon the technical expertise and financial
support of all the municipal corporations in the country. The
interaction with government departments that the Association
enjoys as a means of making known its views, is reinforced by
the influence the Vice-Presidents of the Association can bring
to bear in the House of Commons and the House of Lords. The
approach to the development of hazard control policy is
essentially incremental as the Association deals mainly with
situations at they arise.

POLITICAL PARTIES

A copy of the questionnaire was sent to the three main political
parties, but a response was only obtained from the Conservative
Party Research Department. They described the role they play
rather as that of middle man. They are on the receiving end of
the representations of many organisations, and their evaluation
of all the information they receive is used to inform the party
in general. Their role then is to provide an information
service to Ministers and backbench MPs. That gives a critical
assessment of what may rather loosely be termed the views in
the country. This suggests that the Conservative Party Research
Department is a channel through which interest groups work to
promote their views rather than an interest group working on its
own behalf. Similar organisations in the other two parties
probably perform the same function.

REFERENCES

1. ILERSIC, A.R.   Statistics 13th Edition. Published PFL
                    (Publishers) Ltd. London 1964 pp 298-299

2.   Charters, By-laws, Regulations and Rules
                    of the Institution of Civil Engineers,
                    published by the Institution 1963, p.5

3.   The Annual Report of the Institution of
                    Civil Engineers for 1971 published by the
                    Institution 1972,p.62

4. PARSONS, R.H.   History of the Institution of Mechanical
                    Engineers published by the Institution of
                    Mechanical Engineers, London 1947 pp 10-12

5. TYE, W   Airworthiness and the Air Registration
                    Board. The Journal of the Royal
                    Aeronautical Society Vol.74, No.719
                    November 1970 published by the Royal
                    Aeronautical Society Symposium, February
                    1972, p.2

6. TYE, W   Safety - The Role of the Authority - a
                    paper presented to a Royal Aeronautical
                    Society Symposium, February 1972, p.2

7.   A ticket to Knoxville - published in the
                    Business Section of the Economist dated
                    20 May 1972, p.90

8.   The Charter of Incorporation and The By-
                    Laws of the Royal Aeronautical Society,
                    published by the Royal Aeronautical Society
                    London, 1948 pp 3-6

9.   The 107th Annual Report of the Council.
                    The Aeronautical Journal April 1972,
                    published by the Royal Aeronautical Society
                    pp 217-247

10.   Central Electricity Generating Board,
                    Annual Report and Accounts 1970-71.
                    Published by Her Majesty's Stationery
                    Office pp 31-32

11.   Containment and Siting of Nuclear Power
                    Plants Proceedings of an IAEA Symposium
                    held in Vienna from 3-7 April 1967.
                    Published by the IAEA Vienna 1967 pp 129-170
                    and pp 270-329

12.   Safety and Health at Work. Report of the
                    Committee 1970-72, Vol 2, Selected Written
                    Evidence, Published by Her Majesty's
                    Stationery Office, London 1972 pp 525;543

EXAMPLE OF COMPLETED QUESTIONNAIRE

QUESTIONNAIRE TO ORGANISATIONS ACTIVELY
INTERESTED IN INFLUENCING POLICY RELATED TO THE
CONTROL OF HAZARDS

1. Name of organisation

    SOCIETY OF BRITISH AEROSPACE COMPANIES LTD.

2. How large is your organisation in terms of members and
    employees?

    MEMBERS: ORDINARY   26
                    ASSOCIATE   321
                                   ———
                                   347
    EMPLOYEES: 43

3. In the case of members, are they private individuals or
    corporate bodies?

        CORPORATE BODIES, I.E. AIRFRAME, AEROENGINE AND
        EQUIPMENT MANUFACTURING COMPANIES AND MATERIALS
        PRODUCERS

4. What financial contributions do members make to your
    organisation funds?

        ORDINARY: ANNUAL SUBSCRIPTION OF £3,OOO AND A LEVY
                    BASED ON AEROSPACE TURNOVER

        ASSOCIATE: ANNUAL SUBSCRIPTION AT RATES VARYING FROM
                    £125-£5OO.

5. How is the policy of your organisation formed?

        THE SOCIETY'S COUNCIL IS THE POLICY MAKING BODY AND
        IS SUPPORTED BY VARIOUS COMMITTEES OF A GENERAL OR
        SPECIALIST NATURE

6. What views, if any, has your organisation on the way
    Government policy should be developed on the control of
    hazards?

        THE SOCIETY'S PRIME OBJECT IS TO ENCOURAGE, PROMOTE
        AND PROTECT THE BRITISH AEROSPACE INDUSTRY AND, WITHIN
        THIS BRIEF, PROBLEMS OF AIRWORTHINESS IN MANUFACTURE
        AND MEASURES TO PROMOTE SAFETY IN THE AIR, IN THE
        LIMITED ASPECT OF TEST FLYING OF DEVELOPMENT AIRCRAFT,
        ARE MAJOR MATTERS OF CONCERN. BY ITS REPRESENTATION
        ON GOVERNMENT POLICY COMMITTEES THE SOCIETY, THEREFORE,
        PLAYS ITS PART IN SHAPING THE GENERAL DEVELOPMENT OF
        HAZARD CONTROL.

7. Do you consider that the acceptability of a particular
    hazard can be determined in terms of probability?

        MODERN AIRWORTHINESS CONCEPTS ARE DIRECTLY BASED ON A
        NOTIONAL ASSESSMENT OF PUBLIC TOLERANCE OF ACCIDENTS.
        THE STANDARDS OF SAFETY AND RELIABILITY OF VITAL

EQUIPMENT COMPONENTS AND STRUCTURAL ELEMENTS WHICH
CONTRIBUTE TO THE INTEGRITY OF THE WHOLE AIRCRAFT
ARE DERIVED FROM THESE UNDERLYING ASSUMPTIONS.

8.   On what Government bodies and Committees related to the
control of hazards is your organisation represented?

CIVIL AVIATION AUTHORITY (AIRWORTHINESS DIVISION),
DEPARTMENT OF TRADE AND INDUSTRY (OPERATING DIVISION),
MINISTRY OF DEFENCE (PROCUREMENT EXECUTIVE) - JOINT
AIRWORTHINESS COMMITTEE, AICMA AIRWORTHINESS COMMITTEE
MEETING OF EUROPEAN AIRWORTHINESS AUTHORITIES, UK
FLIGHT SAFETY COMMITTEE, CIVIL AIRCRAFT CONTROL
ADVISORY COMMITTEE AND TECHNICAL CO-OPERATION
COMMITTEE FOR ALL WEATHER CONDITIONS.

9.   What methods, other than those mentioned in reply to the
above question, do you use for influencing government
policy on hazard control?

THE SOCIETY'S STANDING COMMITTEE ON AIRWORTHINESS
REPORTS TO THE TECHNICAL BOARD.   THERE IS ALSO A
JOINT EAA/SBAC ALL WEATHER OPERATIONS COMMITTEE. THE
SOCIETY'S FLIGHT OPERATIONS COMMITTEE ADVISES THE
COUNCIL ON MATTERS AFFECTING THE OPERATIONS OF
COMPANY AIRCRAFT AND THE EMPLOYMENT OF FLIGHT TEST
PERSONNEL.   SUBJECTS COVERED ARE TEST FLYING, FLYING
CONTROL AT DISPLAYS, NATIONAL AIR TRAFFIC CONTROL,
SAFETY IN FLIGHT AND RELEVANT FLYING REGULATIONS.

10.   Do you employ a public relations specialist to promote
your views?

NOT SPECIFICALLY IN RELATION TO AVOIDANCE OF HAZARD.

11.   Can you give details of those cases in the recent past,
where you have attempted to lead or influence Government
policy in relation to hazard control?

THE SOCIETY HAS BEEN REPRESENTED IN DISCUSSIONS ON THE
FOLLOWING SUBJECTS:- REVISIONS AND ALTERATIONS OF AIRWAYS
OVER THE UK, CONTROL ZONES, SPECIAL RULES AREAS,
ARRANGEMENTS FOR HIGH SPEED TEST FLIGHTS, AIRCRAFT CLIMB
CRITERIA, DETERMINATION OF UK TRANSIT LEVELS AND ALTITUDES,
VISUAL FLIGHT RULES CRITERIA, AIRCRAFT ACCIDENT
INVESTIGATION PROCEDURE, AIRMISS INCIDENTS, FLIGHT SAFETY
TRAINING, ALL WEATHER OPERATIONS AND FREQUENT AND REGULAR
PARTICIPATION IN AIRWORTHINESS DISCUSSIONS WITH THE CIVIL
AND MILITARY AIRWORTHINESS AUTHORITIES.

12.   In which of the above cases was your advice accepted and
in what way did you see policy modified:

AIRWORTHINESS REQUIREMENTS ARE CONSTANTLY EVOLVING
AS A RESULT OF THE CONTINUOUS CONTACT WHICH PROMOTES
IDENTITY OF VIEW BETWEEN MANUFACTURERS AND REGULATORY
AUTHORITIES.

13. Is it possible for me to have copies of any reports, such as your Annual Report, that may amplify your answers to the above?

> ANNUAL REPORT 1971/72 WILL BE SENT ON PUBLICATION. BOOKLET "THE SOCIETY OF BRITISH AEROSPACE COMPANIES LTD" IS ATTACHED.

14. Have you any objection to your replies to this questionnaire being published as part of my final analysis of the factors that influence policy on hazard control?

> NO.

BODY	CHAIRMAN
COMPANY LAW	N.P. Biggs
ECONOMIC	Sir Hugh Weeks, CMG
EDUCATION AND TRAINING	H.S. Mullaly
EMPLOYMENT POLICY	L.F. Neal, CBE
ENERGY POLICY	Sir David Barritt
ENVIRONMENTAL AND TECHNICAL LEGISLATION	John Langley
EUROPE STEERING	D.J. Ezra, MBE
FINANCE	Sir Stephen Brown, KBE
GENERAL PURPOSES	John Partridge
INDUSTRIAL RELATIONS AND MANPOWER	N.A. Sloa, QC
INTERNATIONAL LABOUR	C. Henniker-Heaton, CBE
LABOUR AND SOCIAL AFFAIRS	L.F. Neal, CBE
MARKETING	E.R. Nixon
MINERALS	J. Taylor
NEDC LIAISON	John Partridge
OVERSEAS	Sir Arthur Norman, KBE DFC
OVERSEAS INVESTMENT	Sir Duncan Oppenheim
OVERSEAS SCHOLARSHIPS BOARD	Sir Mauric Fiennes
PRODUCTION	J.M. Langham
PUBLIC AND PRIVATE SECTOR RELATIONSHIPS	Sir Richard Way KCB CBE
REGIONAL DEVELOPMENT	W.F. Cartwright TD JP DL
RESEARCH AND TECHNOLOGY	Dr. R.B. Sims
SAFETY, HEALTH AND WELFARE	A.W. Ure
SMALLER FIRMS COUNCIL	R.G. Beldam
SOCIAL SECURITY	R.J. Kerr Muir
STATE INTERVENTION IN PRIVATE INDUSTRY	R.P.H. Yapp
SUPPLIER AND CUSTOMER RELATIONS	D.J. Barron
TAXATION	Alun G. Davies
TRADE PRACTICES POLICY	Sir John Reiss
TRANSPORT	J. MacNaughton Sidey DSO
VALUE AND RATING	J. Taylor
WAGES AND CONDITIONS	A.N.G. Dalton, CBE

LIST OF THE GOVERNMENT AND INDEPENDENT BODIES ON WHICH
THE CONFEDERATION OF BRITISH INDUSTRY IS REPRESENTED

GOVERNMENT BODIES

Cabinet Office
National Economic Development Council
NEDC Committee on Management Education Training and Development,
  and Sub-Committee on Marketing
EDC for the Movement of Exports

Department of Education and Science
Centre for Information on Language Teaching
National Advisory Council on Art Education
National Advisory Council on Education for Industry and
  Commerce
National Committee for the Award of the Certificate in Office
  Studies
Regional Advisory Councils for Further Education

Department of Employment
National Joint Advisory Council and Committees
NJAC Committee on Methods of Payment of Wages
Central Training Council and its Committees
Retail Price Index Advisory Committee
Industrial Health Advisory Committee and its Sub-Committees
Industrial Safety Advisory Council and its Committees
National Advisory Council on the Employment of the Disabled
National Youth Employment Council and its Scottish and Welsh
  Advisory Committees
Expert Group on Earnings Surveys

Department of the Environment
Area Transport Users' Consultative Committees
Central Transport Consultative Committee
Regional Economic Planning Councils
Central Advisory Water Committee
Technical Committee on Discharge of Toxic Solid Wastes
Standing Technical Committee on Synthetic Detergents
Water Pollution Research Laboratory Steering Committee
Technical Committee on Water Quality

Department of Health and Social Security
Industrial Injuries Advisory Council
National Insurance Advisory Committee
National Consultative Council on Recruitment of Nurses and
  Midwives
National Radiological Protection Board Advisory Committee

Department of Trade and Industry
Census of Distribution Advisory Committee
Census of Production Advisory Committee
Standing Advisory Committee on Patents
Trade Marks Advisory Group
Anglo Soviet Joint Commission
Queen's Award to Industry
Iron & Steel Consumer Council
Area Electricity Consultative Councils
Area Gas Consultative Councils
Energy Advisory Council
Industrial Coal Consumers Council
Simplification of International Trade Procedures Board
Foreign and Commonwealth Office
Overseas Labour Consultative Committee
OLCC Aid Sub-Committee

Home Office
Community Relations Commission Advisory Committee on Employment
Crime Prevention Committee
Standing Advisory Committee on Dangerous Substances

Ministry of Defence
Advisory Committee on the Territorial Army and Volunteer Reserve

Ministry of Posts and Telecommunications
Post Office Users' National Council
Post Office Advisory Committees

Scottish Office
Scottish Water Advisory Committee
Schools Industry Liaison Committee
Scottish Economic Council
Highlands and Islands Development Board
Scottish Law Commission
Valuation Advisory Council

Treasury
National Savings Committee

Welsh Office
Welsh Council
Government of Northern Ireland
Northern Ireland Economic Council
Ministry of Development Transport Committee
Northern Ireland Training Council

INDEPENDENT ORGANISATIONS
Advisory Council on Calibration and Measurement
Anglo-Yugoslav Trade Council
Association of British Chambers of Commerce Language Committee
Australian British Trade Association, British Council
British Industrial and Scientific Film Association Council
British Institute of Management Council
Management Consulting Services (Joint Committee)
British National Export Council and Committees
British Productivity Council
British Shippers' Council

British Society for International Understanding
British Standards Institution Executive Council and Committees
British Trade Council in Germany
British Volunteer Programme
British Work Measurement Data Foundation
Business and Industry Advisory Committee to the Organisation
   for Economic Co-operation and Development
Careers Research and Advisory Centre Advisory Panel
Central Fire Liaison Panel
CIRET (International Contact on Business Tendency Surveys):
   Co-ordination Committee
City and Guilds of London Institute,
Advisory Committee on Fuel and Power Subjects
Advisory Committee on Radiation Safety Practice
City Panel on Takeovers and Mergers
Committee on Invisible Exports
Coombe Lodge (Further Education Staff College)
Council for National Academic Awards Business Studies Board
Council of European Industrial Federations
Council of Industrial Federations of EFTA
Council of Industry for Management Education
EFTA Consultative Committee
English Speaking Union Current Affairs Committee
Exhibition Liaison Committee, United Kingdom Committee
Fire Protection Association
Freight Transport Association
Harlow Occupational Health Service
India, Pakistan & Burma Association
Industrial Education and Research Foundation
Institute of Export
International Apprentice Competition UK Committee
International Association for the Exchange of Students in
   Technical Education
International Chamber of Commerce, British National Committee
International Fiscal Association
International Labour Organisation
International Organisation of Employers
Keep Britain Tidy
Metrication Board and Committees
National Council for Quality and Reliability
National Examinations Board for Supervisory Studies
National Institute of Economic and Social Research
National Institute of Industrial Psychology
National Marketing Council
National Reference Library of Science and Invention
River Authorities
Parliamentary and Scientific Committee
Public Schools Appointments Bureau
Royal Society for the Prevention of Accidents
Royal College of Art (Court)
Schools Council for the Curriculum and Examinations and
   Committees
Sino-British Trade Council
Standing Conference on Local Support for Schools Science and
   Technology

Trade Marks, Patents and Designs Federation
United Kingdom Automation Council
United Kingdom/South Africa Trade Association
University Grants Committee
West Africa Committee